# SEASONS OF LIFE

ALSO BY RUSSELL G. FOSTER AND LEON KREITZMAN

*Rhythms of Life*

# OF LIFE

*The biological rhythms that enable living
things to thrive and survive*

Russell G. Foster & Leon Kreitzman

Yale University Press
New Haven and London

Dedicated to the memory of
Professor Eberhard (Ebo) Gwinner
1938–2004
*Founding director Max-Planck-Institut für Ornithologie*

Photography by Dieter Schmidl

'He was not only one of the most influential ornithological researchers
of the second half of the twentieth century, but also a remarkable friend,
colleague and mentor' (Brandstaetter & Krebs, 2004).

# CONTENTS

# FOREWORD

In their first book on biological time, entitled *Rhythms of Life*, Russell Foster and Leon Kreitzman told how cells measure the time of day with what is called a circadian clock. That truly remarkable clock has a periodicity of 24 hours and shows all the hallmarks of a watch devised by us humans. Each day the 'pips' make sure that it keeps precisely in tune with the rotation of the earth and it is temperature compensated. Its role is to ensure that the myriad events of life occur in sequence each day. In this volume – *Seasons of Life* – they take us on to the story of how life adapts to changes of season. Those changes are dramatic indeed at the poles, and it is obvious that without great alterations life in these extreme regions would be impossible. That applies to a slightly lesser extent in the temperate zones but it is still critical to survive cold winters and hot summers, and to produce young at just that season of the year for them to survive to maturity. It comes as no especial surprise, therefore, that animals and plants not only have a daily clock but also possess a calendar that tells the organisms what time of year it is.

We understand much about the calendars involved but not yet enough. As this book shows, there are (at least) two sorts of calendar. One is based on circadian clocks and allows the organism to work out daylength, which is a sound proxy for time of year. If that is not remarkable enough, then the other way of measuring time surely is! Many organisms seem to possess a clock with an innate periodicity of about one year and, by analogy with the daily clock, it is called a circannual clock. How it is constructed is unknown but for animals that hibernate or spend their winters in the deep tropical forests away from changes in daylength, it offers a precise calendar.

That may be how organisms time the year, but this is where seasonality really starts to become intriguing. Virtually every process is best carried out at one time or other of the year and that is not something to take lightly. Survival depends on it, and transmitting one's genes into the next generation certainly does. Much of this book is devoted to the wonderful processes that occur seasonally. The most obvious examples are the migration of birds and large mammals to avoid unpleasant winters and yet who return with unerring accuracy the following year. The alternative strategy is to avoid inclement winters altogether by hibernation in rodents, diapause in insects, dormancy in plants. Finally one has to breed at that time when the young stand the best chance of themselves surviving to maturity. This is invariably determined by the availability of food for the newly born offspring, and so the seasonal cycles of many species become inextricably connected. This is where climate change is intruding into this carefully evolved system. Earlier springs cause caterpillar emergence to advance in Wytham Woods near Oxford, and so the great tits had better track these changes and advance the date of egg-laying or they will become extinct.

There is much more to seasonality, though, than breeding or migration. Virtually every internal process alters. Male red deer restrict food consumption quite drastically during Alpine winters as part of their survival strategy, even though, of course, they do not hibernate. This does not happen because food is absent, although one might imagine so. Take the deer down to lower altitudes and still it eats less.

We humans were clearly highly seasonal beasts until the coming of electric light but traces remain. So-called seasonal affective disorder (SAD) is an example, although as in so many stories one has to disentangle fact from supposition. Foster and Kreitzman attempt this and many other challenges in what is a wonderful story.

Sir Brian Follett FRS, Department of Zoology, University of Oxford

# ACKNOWLEDGEMENTS

Several summers and autumns – and winters and springs for that matter – have passed since we first thought of writing this book. We tried the patience of family, friends and publishers but we managed to finish, with no small thanks to academic colleagues who reviewed our work with a critical eye.

Barbara Helm, Stuart Peirson, Jim Hardie, David Saunders and Daniel Rock read through the entire draft, and their helpful comments on the overall structure and content has been of enormous benefit, as has the particular expertise they brought to various chapters. Harriet McWatters kindly read Chapter 3 and gave us the benefit of her friendly criticism and comment; a very big thank you is due to Barbara Helm, who also advised on Chapters 4, 5, 6 in detail; Andrew Loudon and David Hazelrigg provided expert help on Chapter 4; Bambos Kyriacou reviewed Chapter 5; Till Roenneberg was extremely helpful with Chapters 8 and 9, and Daniel Rock provided valuable comments and material for Chapter 12.

Ben Lobley and Sally Pellow read the entire draft for style and coherence, and the late Mike Karger assisted on early drafts. Philippe Rousseau kindly helped with our references to 'classical' writings.

We would also like to thank the referees who read through the first version of the final manuscript at the request of the publishers. They will recognise their anonymous but vital help in the comments and critiques incorporated into the final text.

Any errors are solely down to us, and the last people to whom they can be attributed are Andrew Franklin of Profile Books & Serpents Tail, and Jean Black of Yale University Press. We are fortunate in having had guidance from such distinguished and wise editors.

Helga Gwinner graciously put us in touch with colleagues who had worked with her late husband, and also provided us with relevant papers and books.

Our thanks go to colleagues at The Nuffield Laboratory of Ophthalmology and Brasenose College, University of Oxford, for their sustained enthusiasm and support for this project.

This is the second book we have written together, and familiarity has its ups and downs. That we managed without too much friction is due in no small part to the support of our families. Thanks are due again from L.K. to Linda, Sophie and Leah, and from R.G.F. to Elizabeth, Charlotte, William and Victoria. Both of us know how boring it can be for the rest of the family when the writer retreats to 'the book', and we are grateful for the forbearances we have been shown.

# INTRODUCTION

*While the earth remaineth, seedtime and harvest, and cold and heat, and summer and winter, and day and night shall not cease.*

GENESIS 8:22

For all living things, each day is a struggle with other organisms and the natural environment. To get through it, an organism has to anticipate change, react to events and expend energy, trying to balance the amount lost with the amount acquired. There is no purpose other than the tautology of survival. These survival days are punctuated and often terminated by an act of reproduction. The one enables the other: survival followed by reproduction. This is the deep, ultimate rhythm of life, the primeval tick–tock of survive/reproduce; survive/reproduce; survive/reproduce. Turtles do it for years, in mayflies it is often over in a day.

Each day, living creatures have to anticipate the predictable rhythms in light and dark and warmth and cold that occur during every 24 hours as a result of the earth's rotation on its axis as it travels around the sun. Plants, animals, algae and even bacteria need to 'know' – in a manner of speaking that must not be taken literally – the time of day.

They can anticipate the predictable daily events because of the ubiquitous internal circadian clocks that enable living things to coordinate their actions with the external world. They can mesh the time within with the time without. These clocks time the circadian rhythms that have a role in many processes: timing when flowering plants open their petals in the morning and close them at night; confining the emergence of fruit flies from their pupae to the cooler dawn hour; controlling the release of fungal spores to maximise their reproductive success. Inject *Escherichia coli* endotoxin into a mouse, and death or survival can be made experimentally a function of the time at which the agent is injected (Haus *et al.*, 1974). In humans, the clocks are responsible for our core body temperature's rising in the day and falling at night; they determine the sleep–wake cycle and control many major regulatory functions that affect our physiology, emotions, cognitive ability and behaviour. Disruptions of circadian rhythms in ourselves are linked to jet lag, mental illness, sleep disorders and even some forms of cancer. It is hard to overstate the importance of these rhythms as a key adaptation to life on earth.

In our previous book, *Rhythms of Life*, we tried to explain the nature of circadian rhythms. But the earth not only rotates, it also travels around the sun about once every 365 days. And it does so with its axis tilted at an angle of 23.5° to the vertical. The result is the annual change in light, heat, rainfall and humidity that we know as the seasons. Few places on the planet have the same weather conditions throughout the year. In the higher latitudes there are at least four seasons, and temperatures can vary from +35°C in high summer to −35°C in the middle of winter. Nearer the poles, the midnight sun means near enough six months of daylight and six months of darkness. Even in the equatorial areas, where daylength is always roughly twelve hours and temperatures may not vary much throughout the year, there is often a dry season and a wet season.

We are all familiar with the changes that come with the seasons. As the days become shorter and colder, birds in the higher latitudes migrate to a warmer climate, plants die back and hibernating animals go into their winter sleep. With the lengthening days and rising temperature in spring, the birds return, plants bud and flower, small mammals arouse from hibernation, insects emerge from their pupae, bees buzz and for many animals it is the time for birth.

As well as daily clocks, living creatures need calendars; they have to know the time of year if they are to feed and breed. Even those that do not live through a complete seasonal cycle have life histories that enable them to maximise reproduction. And those organisms that reproduce more than once in a year still generally try to fit their breeding into a specific time slot.

Nearly all the earth's organisms live in a very complex temporal environment. Although a meadow, for instance, is a single environment when we think in spatial terms, it is far more complicated when it comes to time. The meadow by day is not the same meadow at night. It changes throughout the 24 hours. And the meadow is a different environment in each of the seasons. A meadow-living rabbit has to cope with all this temporal change if it is to feed, survive and reproduce. No wonder it spends so much time in its burrow!

Although circadian rhythms can be studied over the course of several days, annual rhythms need a lot more patience. It takes eight years to get eight points on a graph. Yet birds kept in a constant laboratory environment for years still show an annual rhythm of migratory behaviour. Their off-spring, born and bred in these constant conditions, show the same timing capability. Similarly, groups of ground squirrels kept in constant light conditions but at different temperatures – some at near 0°C and some at 22°C – still show hibernatory activity at about the same time of the year. Animals have internal clocks with a period of about one year, called circannual clocks. However, under constant conditions and with no cues to lock the rhythms into synchrony with the movement of the earth around the sun, the circannual rhythms start to drift in time, just as circadian rhythms also drift. But there is little idea as to how these circannual rhythms are generated, how they keep regular time, and how they transmit the timing signal to the rest of the organism.

Many scientists have been fascinated by the ability of plants and animals to sense both the time of day and the time of year and to use the information to synchronise their physiology and behaviour so as to maximise their chances of survival and reproduction. Ebo Gwinner was one of those who spent his life trying to find the answer to these questions. He was a naturalist at heart with a passionate interest in birds, and he was fittingly the founding director of the Max Planck Institute for Ornithology in Bavaria. Gwinner pioneered studies of circannual rhythms in birds and their role in organising migration, breeding and other life-history events. Gwinner was a kind and modest man, and this book is devoted to his memory and the example he

3

set of patient, determined study of one of the natural world's many wonders.

*Seasons of Life* tries to explain how living creatures know that the seasonal change is coming, how they prepare for it and how important this is to all of us. These seasonal changes are difficult to study, but the science is progressing quickly. We have tried to give an accurate and up-to-date account, and although we have concentrated on key areas we have had to leave a lot out. Omission is not in any way meant to devalue the worth of those whom we have not mentioned, be they scientists or species. It is merely a reflection of the breadth of the topic and the amount of material.

We have also had to weigh how much detail to include even if it means that in places the text is a tough read. This is in part because for many readers words such as arylalkylamine *N*-acetyl transferase do not trip lightly off the tongue. We have tried to be careful not to use terminology as a form of shorthand and also not to expect an easy familiarity with biological concepts. For the lay reader, who may find some of the terminology somewhat daunting, we suggest that the key is to try to grasp what is happening, which is really quite simple, rather than getting stuck in the detail.

The first chapter explains why seasons occur. It is surprising, given how sophisticated we in the developed world are supposed to be, how few of us fully understand why the seasons happen.

Chapter 2 concerns itself with what might be called the natural history of the seasons. Peoples who are far closer to the land than most of us are adept at reading the natural signs that predicate seasonal change, and they count their calendar not just from the skies but also from the ever-shifting relationship between plants and animals. This close affinity to nature has been lost by most of us in the developed world, although there are still a handful of those who can tell the species of a tree by the sound it makes as the wind blows through.

In Chapter 3 we move into the first of the four sections detailing the mechanisms used by plants and animals to anticipate seasonal change, and we start with plants. Although few of the readers of this book will be farmers, many will be gardeners. They know that some plants, such as camellias, flower early in the year and some, for example chrysanthemums, flower later and that getting the timing right for germination, bud burst, flowering and a wide variety of other features is critical if plants are to survive and success-fully reproduce. While the seasonal nature of plants was recognised way back in antiquity, it took all of the twentieth century to work out how it was done – and although we know a great deal, we still do not have the whole picture. In the course of this chapter we introduce photoperiodism, which simply means the ability of an organism to measure and respond to the length of a period of light (photoperiod), or conversely a period of darkness. This is the key concept and a very important basic biological principle.

Animals aim at producing their young when there is plenty of food available – which of course ultimately depends on the plants. In Chapter 4 we concentrate on how mammals and birds maximise the life chances of their offspring by using a photoperiodic response, similar in principle to that in plants, as a way of synchronising their life cycles to seasonal change, so as to optimise reproduction.

The idea of a deep, death-like sleep has long fascinated humans even though we do not hibernate. In fact, mammals over about five kilograms in body weight are not true hibernators. Despite this, notions of suspended animation have been a staple of science fiction writers. Chapter 5 considers dormancy and hibernation in a wide range of plants and animals.

Not all organisms tough out harsh conditions by staying put. Avian mass migration may be one of nature's most spectacular shows, but many other animals make a solitary round trip every year. Chapter 6 recounts how animals use timing mechanisms, including an endogenous circannual clock, to signal when to migrate on their often long journeys.

The last section of the book is largely given to exploring the effect of seasonal change on humans. We are reflexive creatures, and the seasons play a large part in human culture and in making our psychology and social be-haviour what it is. Chapter 7 documents the important and probably crucial role of seasonal change in forming us physically, psychologically and so-cially. The story starts in Africa when our early ancestors began the evolu-tionary journey that has lasted right up to the present day. The adaptations of our ancestors that enabled them to survive and reproduce by anticipating the seasonal vagaries of the climate still reside deep within our metabolism and life histories.

Chapter 8 is about conception and birth. There is a seasonal gradient to human births, which tend to peak in early spring in the northern hemisphere. Although seasonal timing of births might be expected in societies that live close to the natural world, it persists even in today's modern world in which '24/7' lighting and constant food largely masks us from seasonal change.

Humans are not seasonal breeders in the same sense as other mammals such as sheep and hamsters, which time the arrival of their young to the point of maximal food availability. Humans, just as chimpanzees and gorillas, are ready to procreate more or less at the drop of a hat, more or less most of the time, provided that the female is neither pregnant nor lactating. So there is a puzzle: how and why do these largely opportunistic breeders show marked birth seasonality?

The answer seems to be that the timing of human birth may depend more on the nutritional status of the mother at conception than the availability of food at the time of birth. This matters because in Chapter 9 we look at some of the relationships between the month of birth and the pattern of disease in later life – a phenomenon unsurprisingly known as the 'month of birth' effect. How long you live, how tall you are, how well you do at school and how likely you are to develop a range of diseases, including devastating conditions such as schizophrenia, may all be modulated to some extent by the time of year in which you emerged from the womb.

In Chapter 10 we explore some of the work that is going on in understanding the relationship between illness and seasonal factors, and in particular between our internal physiology and susceptibility to external agents of disease, whatever their source. Our moods, performance, sleep patterns, thermoregulation capability, thyroid function, cortisol levels and almost everything else one cares to mention show some seasonal variation. And perhaps the best-known instance, which we discuss in Chapter 11, is seasonal affective disorder (SAD).

Since it was first recognised in the early 1980s, there has been something of a battle over whether SAD is in fact a discrete condition rather than just a depressive state that happens to coincide with certain times of year. However, the latest evidence is that SAD is a real depressive condition that is seasonally triggered. But instead of thinking of SAD as a specific disorder possibly caused by light deprivation alone, it seems that we should consider it as a complex disorder with genes, environment and culture contributing to its aetiology.

The seasons play a large part in forcing the timing of our deaths on us, and Chapter 12 explains that the relationship between the seasons and time of death is neither simple nor straightforward. Over the past couple of centuries, there has been a decline in mortality from diseases that had previously been most severe in the summer months. These declines resulted from rising living standards, improved nutrition and food storage, cleaner water supplies and better sanitation.

Over the same period there has been an increase in mortality from diseases associated with colder months, such as diseases of the circulatory system and respiratory diseases such as pneumonia and influenza. Whether these patterns of mortality and morbidity will continue much as they are, with the changes in our climate, is a moot point.

In Chapter 13 we consider some of the issues raised by the changes in the timing of the seasonal climate. It is clear that species are already being affected, and the delicate temporal web that has been painstakingly built by the agency of natural selection operating on small variations over millennia is being disrupted as a result, most probably, of human behaviour.

As the physical environment changes with a change in climate, many species will extend their range across the earth's surface. Others may live at a different altitude. Some will do both. Some will adapt to the changing world and will be able to accommodate the new timings needed for seasonal prediction. However, many will not and will become extinct.

Understanding the ways in which organisms have adapted and will adapt to the timing of seasonal climatic change will help us in the future to both mitigate and manage some of the effects of global climate change. A better understanding of the biological processes will help us to develop new agricultural and horticultural practices, and to devise methods for protecting human health from attacks by both old and resurgent pathogens and new and insurgent ones. It will also enable us to try to preserve and conserve other species.

# 1

# THE GENERATION OF THE SEASONS

'The essential joy of life is seeing seasons unfolding'
MARY CASE, A BEEKEEPER ON THE ISLE OF WIGHT (CASE, 2008)

Without the sun there would be no life as we know it. Billions of years ago, sunlight powered the first photosynthetic bacteria that provided the oxygen in our atmosphere and later powered the plants that in turn became the fossil fuels. Today sunlight still powers all the photosynthesising organisms on the planet. The sun drives our weather systems and provides us with the means of life. The energy figures are mind-blowing: in just over one millionth of a second, the sun radiates as much energy as humanity produces in a year. If we could continually harvest the energy coming from just 100 metres squared (10 hectares) of the surface of the sun, we would have enough to supply the current daily energy demands of the world.

During all the time that the sun has been shining on it, the surface of the earth has constantly changed. The tectonic plates moved, mountains rose up and islands sank, continents formed and reformed and became warmer or colder, volcanoes came and went, lakes became dry land, and forests turned into deserts and vice versa. Along with the physical changes at the surface, air streams have altered course and ocean currents changed direction. Climate change itself is not new. Throughout the billions of years since the earth–moon duo stabilised their orbits, the earth's rotation on its axis slowed to its present 24 hours or so while it made its annual 365 and a bit days' journey around the sun. During the course of the year the days lengthened and shortened with metronomic accuracy, and with these changes came seasons.

Seasons as we know them happen because the amount of energy at a given point on the earth at different times of the year changes with the angle and hence the intensity of sunlight falling on the surface. This effect is compounded by the duration of the day: the farther from the equator, the greater the change.

The annual cycling in the energy received results from the earth's spatial relationship with the sun. Seasons are determined by the fundamental features of the planet – its orbit round the sun; its geometry in space; the disposition of the land and water. Seasons are not chance events, but they come and go across the face of the globe in an orderly manner. Whether it is the arrival of snow in winter, monsoon rains, or summer heat, our environment has been changing constantly but predictably every year, and these profound changes occur over relatively short time periods. This seasonal variation shapes the life cycles of animals and plants as they struggle to cope with the changes in their local environment. Their adaptations to this regular change play a large part in determining their survival and reproductive success.

The same goes for us humans. The difference is that we have developed and adapted in such a way that we survive seasonal change by modifying the environment in which we live: through shelter, clothing, and now air conditioning and central heating. We have shaped the environment to our needs. It is this capability that accounts for the enormous geographic range of humans. It is true that beavers build lodges in which they see out the winter, bees have hives, termites have their mounds and many animals live part or even all of their lives in burrows, but in general plants and animals modify themselves through their life cycle, adaptations and behaviour patterns rather than changing the immediate environment in which they live. Non-human primates, for instance, are restricted to environments to which they are naturally adapted. Great apes (chimpanzees, gorillas, bonobos) live 20° north or south of the equator and there are no primates nowadays indigenous to Europe or North America.

But in modifying the world, we have lost contact with nature and its timing. In the last few centuries, as we have moved en masse from the land to the city, we have become detached from the close association with the seasons. We are a long way from the Squamish people, aboriginal to British Colombia, who believed that the singing of a thrush (*Hylocichla ustulata*) is responsible for ripening the salmonberries (*Rubus spectabilis*). They had similar markers for all times of the year, whereby one natural event signalled to them the advent of another. Their calendar was all around them – in the trees, in the sky, in the rivers and on the ground, as long as one knew the code.

This deep understanding of the dynamics of their local environment was true for all indigenous peoples. When the French explorer Samuel de Champlain arrived at Cape Cod in 1605, the Wampanoag people informed him that the best time to plant corn was when the white oak (*Quercus alba*) leaf was the same size as the footprint of a red squirrel (*Tamiasciurus hudsonicus*) (Lantz & Turner, 2003). A hemisphere and several hundred years away, Frances Bodkin, a descendant of Australian D'harawal Aborigines and a botanist at Sydney's Mount Annan Botanical Gardens, is rightly impressed by the knowledge of his forebears (Reuters, 2003):

> Present-day scientists do their studies by measurements and experiments. Aboriginal people are just as good scientists, but they use observation and experience. In 1788, when English settlers first arrived in Sydney, they imposed the four European seasons on their new home without any real knowledge of local weather patterns, yet the local Aborigines lived according to an annual six-season calendar, based on the flowering of various native plants.

The transitional points in an animal or plant's life cycle that are linked to seasonal change are known as phenological events. The ritualistic notification to the London *Times* newspaper of the first cuckoo of spring is acknowledging a phenological event, as is the first flowering of daffodils. Phenology is in essence the study of the times of recurring natural phenomena. Trevor Lantz and Nancy Turner of the University of Victoria have pointed out (Lantz & Turner, 2003):

> The arrival of one event predicts the imminence of another. This data is a valuable predictive tool in forestry, agriculture, and fisheries. Fishermen in Western

Canada have long recognized that pickerel (*Essox lucius*) run at the time when the southern cottonwood (*Populus balsamifera*) releases seed, and on the East Coast of Canada, fishermen would not fish for shad (*Alosa sapidissima*) until the saskatoon, or shadbush (*Amelanchier*), had flowered.

Indigenous people know that seasonal events happen with predictable regularity. But it would not surprise us that they did not know why. The best that the Quechua people of the Andes could do was their belief that the sun shrank when it became thirsty (during the dry season) and became bloated when it gulped river water (during the rainy season). This is very charming in its own way and typical of the explanations found in many cultures, but hopelessly wrong.

We would expect university students to do better. Boston, on the east coast of the USA, is at latitude 42° N. In the winter depths, the mean daytime temperature is around freezing, although records show that it has dropped as low as −28°C. Mid-summer days are balmy if a little humid and usually in the mid-twenties Celsius, with the occasional very hot day in the high thirties thrown in. This seasonal variation is familiar to everyone who lives there, including Harvard students.

But when twenty-five new Harvard graduates at a Commencement Day ceremony were asked a simple question, 'Why is it hotter in summer than winter?', only three got it right. Twenty-two out of twenty-five, including some who had majored in a science, could not give the correct answer. They did not know why the seasons happen (Schneps & Sadler, 1987). Harvard graduates are an easy target, and in fairness they are hardly alone in being so ignorant of the world around them. Most people seem not to know why the seasons happen. In the main, they think that the earth is nearer the sun in summer than in winter. But at present the earth's orbit around the sun is nearly, but not quite, circular. At its nearest, in the first few days of January, it is 147 million kilometres from the sun and at its furthest, in the first week of July, it is 152 million kilometres away. Averaged over the globe, sunlight falling on the earth in July is a little less intense than it is in January but this has a small effect on temperature.

Or they answer that the sun is hotter in summer than winter. The sun's heat output does vary, and even small changes can make a large difference.

A 0.3 per cent drop in its output in the sixteenth century caused a mini-Ice Age in the North Atlantic region, and London's River Thames froze. But there is no regular change in the sun's size or output that corresponds to the annual cycle of the seasons.

It is all part of a wider picture of a lack of understanding of science in general. Although Copernicus's *De Revolutionibus Orbium Coelestium* was published just before his death in 1543, getting on for 500 years later nearly one-quarter of European adults surveyed by the European Union still think that the sun goes around the earth. Depressingly, more than half of those in the same survey classified astrology as a science (European Union, 2001).

This ignorance of the causes of the seasons is a pity because much of the natural world, including ourselves to a much greater extent than we imagine, is shaped by the variation in weather that accompanies the changes in daylength that signal where we are in the solar year.

The big clue to understanding the generation of the seasons is that when it is summer in the northern hemisphere, it is winter in the south and vice versa. Because both hemispheres are on the same planet and so are the same distance from the sun, a moment or two's thought should suggest that distance from the sun has little to do with it. Another clue is the change in daylength throughout the year as we move north or south of the equator. What matters is the 23.43° tilt of the earth's axis. This tilt from the vertical (to be more exact, to the plane circumscribed by the earth's orbit around the sun), just a few degrees less than that needed to topple a London double-decker bus (Avison, 2000), accounts for our seasons, the distribution of plants and animals and – in the end – us.

Let us start by imagining that from somewhere in space we are looking at the earth orbiting the sun. If there were no tilt to the earth's axis, then as it rotated once every 24 hours on its current near-circular orbit, everywhere on earth would receive 12 hours of sunlight and 12 hours of darkness during each earthly rotation. The sun would rise and set at the same time each day. There would still be variation in temperature, from hottest at the equator and coolest at the poles because of differences in the angle of incidence of sunlight at different latitudes. The sunlight is most concentrated at the point at which the rays of the sun are perpendicular to the tangent of the earth. As the angle between the sun's rays and the tangent becomes smaller,

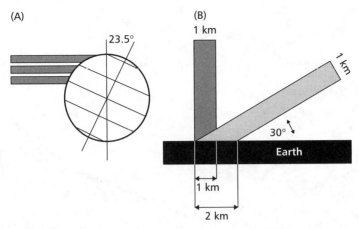

**Figure 1.1** (A) The earth tilts by 23.5° on its axis. The farther from the equator one goes, the smaller is the angle of incidence at which rays from the sun strike the earth. The same amount of solar energy spreads over a larger area, so there is less energy per unit area. (B) A column of sunlight striking the earth perpendicularly to its surface will result in a column-wide area of illumination. But if the incidence of the column of light is 30° to the surface of the earth, the area of illumination will be doubled and the energy per unit area halved.

the effect is that the same amount of energy is spread over a larger area, so each unit of area gets less energy. There would be weather but not as we know it, and there would be no seasons (Figure 1.1).

But the earth's imaginary axis drawn through the poles does tilt, and this makes all the difference. The tilt probably came about nearly four and a half billion years ago. Before the planets finally coalesced into their current line-up, there were asteroids, planetoids and planets and maybe the odd comet barging into each other like an ill-regulated dodgem track. The nascent earth was probably in a monster collision with another would-be planet about the size of Mars. The resulting crash knocked the earth off its axis. It must have been quite a bang. The force 9 earthquake on 26 December 2004 that triggered the killer tsunami off Indonesia may have caused the earth to tilt just one or perhaps two centimetres further off kilter.

As the earth moves around the sun, its axis points towards Polaris, the pole star, in the North and towards the Southern Cross in the South. If the northern hemisphere is tilted towards the sun, the northern hemisphere receives the more direct rays of the sun (that is, the angle of incidence is higher) and it is summer in the northern hemisphere. If the southern hemisphere is

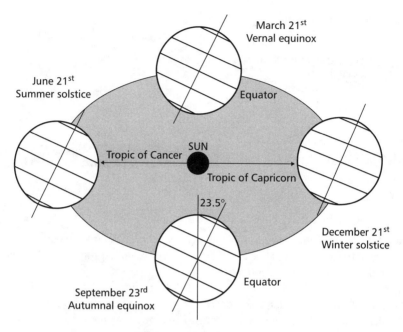

**Figure 1.2** The sun's rays are perpendicular to the Tropic of Cancer, which is 23.5° N, at the June solstice and likewise at 23.5° S at the Tropic of Capricorn at the December solstice. These latitudes are the boundaries of the Tropics, and outside them the sun does not rise high enough in the sky for its rays to be directly overhead. In the northern hemisphere the sun's path moves across the southern skies and vice versa in the southern hemisphere. On the June solstice, the sun is directly overhead at the Tropic of Cancer. It is light at the North Pole, but dark winter at the South Pole. On this day, every location north of the Arctic Circle is illuminated for the entire 24 hours. As it happens, the amount of solar radiation reaching the earth's surface actually peaks at 30° N instead of 23.5° N. This is because sinking air at this latitude (caused by global circulation patterns) produces drier climates, resulting in an abundance of clear days that allow more solar radiation to reach the earth's surface. On the other half of the globe, the circle of illumination never reaches the South Pole, and in fact it does not get to within 23.5° of the South Pole. So every place more than 66.5° south of the equator misses out on sunlight entirely that day – they get 24 hours of darkness. The positions are reversed at the December solstice: the Antarctic is illuminated for the entire 24 hours, whereas north of the Arctic Circle there is 24 hours of darkness (Press *et al.*, 2003).

tilted towards the sun, the southern hemisphere receives the more direct rays of the sun and it is summer in the southern hemisphere. Depending on latitude, the sun appears to climb from low on the horizon in winter to overhead in high summer. This movement of the sun means that the angle at which sunlight hits the earth at a given point changes during the year.

The tilt of the earth on its axis and its daily rotation determine the actual amount of energy (insolation) falling at any given place on the earth at any given time on its annual journey around the sun. This change over the year is vividly seen near the poles, when the summer day is 24 hours of light and the winter day is 24 hours of darkness (Figure 1.2).

If the earth had no atmosphere, the entire energy received from the sun would reach the surface undisturbed. But as the wall of energy moves onwards, roughly one-quarter is reflected back into space by clouds and atmospheric particles. One-fifth is captured by simple absorption by the earth's atmosphere. This interaction with the atmosphere reduces the solar energy input at the surface by about 30 per cent on a very clear day to nearly 90 per cent on a very cloudy day.

Even then, only part of the solar radiation reaching the ground or ocean surface is absorbed. The proportion reflected back upwards depends on what is known as the albedo. This is a measure of the reflectance of the surface. The ice at the poles reflects back about 90 per cent of the energy that falls there, whereas less than 10 per cent is reflected back by water. When sea ice melts, the albedo at the melt-edge changes from 90 per cent to 10 per cent, and the absorption alters from 10 per cent to 90 per cent. The extra heat absorbed causes further melting, setting up a positive feedback. It is this and other positive feedbacks that make climate change so scary and why we are now talking of tipping points. Furthermore, the earth also radiates infrared heat back into the atmosphere. Some of this energy, which has a different wavelength from that coming from the sun, is absorbed by gases in the atmosphere.

If there were no greenhouse gases, which in the main are water vapour, carbon dioxide and methane, much of the heat that is reflected or radiated back would be lost and the average temperature on the earth's surface would be well below freezing. The atmosphere that sustains us is largely a product of life itself. The biological and physical systems of the planet have evolved into a passive control system with many feedback loops that keeps the climatic systems within bounds. With a life-sustained atmosphere, the earth's temperature averages 13°C overall and about 16°C on land.

The weather system is best thought of as an engine whose energy is supplied by the sun. The earth's gravity, curvature and daily rotation combine to ensure that the atmosphere can only move within certain limits. The actual weather results from the air flows set up by the variation of temperature and pressure across the globe. Air warmed by the sun near the equator rises and flows towards earth's poles, returning to the surface and flowing back to the equator. This motion, combined with the earth's rotation, moves heat and moisture around the planet, creating winds, currents and weather patterns. The intensity of the energy falling on a given point (in other words the amount of energy per square metre; Figure 1.1) depends on the angle at which it hits the earth, and the daylength in turn determines its quantity. The changing amount of energy through the year is reflected in changing weather patterns that we know as the seasons.

For plants, animals and us, what really matters is what happens locally. For instance, in summer, air over Asia is heated by the sun, rises, and draws moist air from the Indian Ocean, causing daily rains over most of southern Asia. In winter, the air over Asia cools, sinks, and flows out, pushing the moist air away and creating dry weather. A similar pattern occurs in the Pacific Ocean near Mexico and brings moist air and afternoon thunderstorms to the southwestern United States in the summer.

Ocean currents are driven by the winds and follow the same general pattern. The continents block the flow of water around the globe, so ocean currents flow west near the equator, then turn towards the poles when they strike a continent, turn east, then flow back to the equator on the other side. In all the oceans, the currents form great loops that flow clockwise north of the equator and anticlockwise south of it.

Water in the polar regions is very cold, salty, and dense. It sinks and flows along the sea floor towards the equator. Eventually, the water rises along the margins of the continents and merges with the surface water flow. When it reaches the polar regions, it sinks again. This three-dimensional movement of water mixes heat throughout the oceans, warming polar waters. It also brings nutrients up from the deep ocean to the surface on a seasonal basis, where they are available for marine plants and animals (Press *et al.*, 2003).

Not all the environments on the earth change dramatically or even moderately with the seasons. In some natural environments, such as at great sea depths or in deep caves, the temperature is constant and there is little or no diurnal or seasonal change in the key variables. The aquatic environment is generally more stable than that on land, but depending on the depth and closeness to the shore, there will still be regular changes in factors such as daylength, temperature and salinity.

On land some burrowing animals, such as the naked mole rats that live underground in the deserts of Africa and the Middle East, create their own stable temperature environments. Others modify their environment, but most animals, plants, algae and micro-organisms have had to adapt to whatever nature throws at them and exploit it as best they can.

In deserts, the daily changes in temperature between day and night can be greater than the overall changes due to seasonal variation. Organisms that live in extreme environments are specially adapted to deal with the cold, or the heat, or the wet, or the wind. Most organisms have to cope with far less severe changes. In the UK, the warm Gulf Stream helps to moderate the climate so that the yearly temperature range from high to low is usually about 25°C. In contrast, the winter temperature at similar latitudes in Siberia and Canada, for example, can fall below −40°C, whereas in summer it can rise above 30°C, giving an annual range of about 70°C. In general, the seasonal difference in the temperature at the earth's surface (actually it is measured at 1.6 metres above the surface) tends to be greatest in higher latitudes, and particularly in the interior of large landmasses such as North America and Asia, far from the moderating influence of the oceans. The smallest annual temperature range occurs in the tropical climate zone (Bridgman & Oliver, 2006).

In the long days of summer, there is more time for the sun's rays to heat up the earth, and this is a contributory factor to why it is hotter in summer than in winter. The summer solstice, the longest day of the year, is seldom the hottest. This is due to heat capacity, the physical principle that makes fire-walking possible. The intrepid souls walk briskly across a bed of ash-covered coals without burning their feet. Ask the fire-walker to cross a sheet of heated copper and you would see a different result!

There is more water in the southern hemisphere than the northern. In global terms, January is the coolest month because that is when our planet presents its water-dominated hemisphere to the sun. We are a little closer to the sun in January, but the extra sunlight gets spread across the oceans.

Southern summer in January (perihelion) is cooler than northern summer in July (aphelion) because there is more land in the north and more water in the south. Land warms up more quickly than water because land has a lower heat capacity than water. However, it also cools more quickly. The ocean warms up slowly but it loses heat slowly and so keeps the surface air temperatures up. During summer, the greater amount of land in the northern hemisphere gets heated up quickly, but it loses much of that heat at night and so it takes weeks of warm weather to build up the surface temperatures (Bridgman & Oliver, 2006).

Most living things anticipate the predictable seasonal change in weather. Some animals run or fly away from adverse change and may migrate thousands of kilometres; others stay put, lower their metabolic rate and wait it out. Some put on winter clothing. As the days shorten in autumn and their prolactin levels drop, mink (*Mustela vison*) grow the thick winter coat (Martinet *et al.*, 1984) so much prized by a certain type of shopper (O'Reilly, 1980). Plants time their life-cycle stages to seasonal activity and insects do the same, such as the emergence of the butterfly from its protective chrysalis in late spring or early summer. Nearly all plants and animals time their reproduction to seasonal cycles in weather and resources. Many birds exploit the springtime burst in food in the high latitudes of the Arctic to feed their young. Whales migrate up the western seaboard of the Americas to feed on the plankton blooms in the cold northern Canadian waters. Other animals, such as the meerkats of the Kalahari, stay put and survive, exploiting seasonal change by modifying their behaviour (Clutton-Brock *et al.*, 2001).

Although few natural phenomena change so radically and unpredictably as the daily weather, there is predictability to the seasons. There will be wet periods and dry periods; months that are hotter or colder than others; some months that are windy and others calm. These periods come round at more or less the same time each year and although neither we nor plants and animals can predict the weather on a given day a year hence, all of us have a good idea of the outer limits. To put it another way, climate is what you expect, weather is what you get.

The weather is chaotic in the sense that it is sensitive to initial conditions, but seasonal change is driven by the monotonic regularity of the earth's orbit around the sun. The amount of sunlight falling on a given part of the earth is calculable years in advance and also in the past, and the Serbian mathematician Milutin Milankovitch did just that. How much of it reaches the earth's surface is another matter and is subject to cloud cover and local weather conditions at the time. Furthermore, the effect it has on the weather will be mitigated by the rate at which heat gained is then radiated back into space, which depends on the topography, vegetation, albedo and dozens of other factors including the presence of 'greenhouse' gases. But this does not detract from the overall result: the seasonal climate is variable but in broad terms predictable. Although the theoretical limit for weather forecasting is no more than two weeks (Bridgman & Oliver, 2006), we can have a reasonable assurance as to overall climate a year from now (or at least we used to, before the advent of climate change). There are neither icebergs in the tropics nor sands in the Arctic; nor do we expect them.

Prediction, or perhaps more accurately biological anticipation, is the key theme of this book. In the struggle for reproductive success, organisms have to get their prediction of seasonal timing right. Perhaps the ultimate predictive challenge is long-distance migration. Birds leaving point A for point B have to forecast the climate at point B, which may be thousands of kilometres away, at the time of their arrival. In their breeding grounds in the higher latitudes, migrating birds use an internal calendar to predict their departure time in early autumn and get ready for the migration by laying down fat, shrinking gonads and adjusting their physiology and social behaviour. The birds go through the same process when making the return trip. Although some species arrive and depart en masse, most time their migration departures and arrivals over the course of a few days or even weeks. Their internal predictors that forecast the timing of their journeys is such that *on average* the local weather will be suitable at their arrival. In other words, spring or autumn this year, depending on the direction of travel, has to be only approximately similar to spring or autumn last year for the birds to be all right. If it is very different, they are in trouble. If the springtime earth has dried out sooner than usual, the worms will burrow deeper and worm-eating birds will starve and individuals will die. But even in a difficult year, enough of the population will survive to return the following year when conditions may be more usual. This variation in timing among different subsets in the population enables some bird species to adapt to changes in seasonal timing that can happen even without human interference. Other species may be more 'plastic' in their behaviour and can respond quickly to changing conditions, perhaps changing their destination. It rather depends on how far the birds migrate, but there is an inherent flexibility that has developed within some birds that enables them to cope better with the flexible nature of the environment. Some will always survive until the following year, but what really messes things up for the birds is a sudden and consistent change in the timing of the seasons that is outside the limits of the variation that they can handle.

The predictability of the seasons is vital to most plants, birds and other living creatures including humans, and especially farmers. In his *A Midsummer Night's Dream*, Shakespeare has Titania, the Queen of the Fairies, describe how her argument with Oberon has resulted in drastic changes in the weather:

> The seasons alter: hoary-headed frosts
> Fall in the fresh lap of the crimson rose,
> And on old Hiems' thin and icy crown
> An odorous chaplet of sweet summer buds
> Is, as in mockery, set, The spring, the summer,
> The childing autumn, angry winter change
> Their wonted liveries; and the mazèd world,
> By their increase, now knows not which is which.

Humans and other species have adapted to cope with and even take advantage of predictable change, but the unpredictable change described by Titania has always filled our ancestors with dread. It is not the cold of winter itself that is necessarily the problem, but unseasonable happenings such as the late, snap frost. Because such events were universally attributed to the spirit world, virtually all cultures have festivals celebrating seasonal events as a means of placating the spirits or expressing relief that the weather had behaved predictably. Planting and harvesting festivals are common, as are celebrations of midsummer and midwinter. The timing of these festivals and their association with natural markers such as the solstice and equinox tied them intimately to the development of calendars.

The four seasons of spring, summer, autumn and winter have enormous cultural significance in the higher latitudes, but they are not the only way of dividing up the meteorological year. Whereas the occasional visitor to the far north observes two seasons, the long-day summer and the long-night winter, Inuits know more than a dozen seasons, all signified by animal and astronomical events. The !Kung, the indigenous people of the Kalahari, who have lived there for at least 20,000 years, identify five seasons, although the climate of the central Kalahari is only a two-phase cycle of summer rains and winter dryness (Kolbert, 2005). The !Kung have a keen awareness of the natural world, and their existence depends on a profound understanding of the relationship between animals and plants and the local environment. During the rainy season the seeds dispersed over the desert floor germinate and the plant life flourishes, animals of all kinds gestate and lay eggs, and water holes fill. Families come together in a larger group because the concentration of people can be supported in one area. The group travels only a few kilometres foraging for food. As the season begins to dry, the people disperse. The concentrated group breaks up and the bands go to different water holes.

Although many of the animals in the region migrate, others tough it out and are adapted to life in the harsh climate. Animals in the Kalahari have a lower metabolic rate than their counterparts in other parts of the world. This adaptation allows animals to survive with less food and water. Even so, a meerkat (*Suricata suricatta*) living in the Kalahari loses 5 per cent of its body weight overnight, making the search for food very important every day (Russell *et al.*, 2002).

Prediction enables living creatures to survive the changes in weather and to reproduce at the optimal time for success. Living organisms can handle change. Every 24 hours all creatures on the planet experience cyclic changes in their environment. They can deal with the myriad changes in light and temperature, ultraviolet flux, humidity and nutrient gradients because despite the extent of the changes there is a predictability to them. Day will follow night and organisms have internal circadian clocks that help them coordinate their activities to local time. They can also deal with the seasons because they can anticipate when they are coming.

The length of the day, known as the photoperiod, is the most reliable indicator of seasonal timing. Many plants and animals continually measure the length of this photoperiodic signal and use the information both to synchronise to environmental conditions and to anticipate them. However, because the photoperiod is not rigidly correlated to local temperature, organisms may use the photoperiod combined with other cues, such as temperature, to fine-tune the timing of events. Further, not all living creatures use photoperiod and not all anticipate seasonal change. Many simply react to short-term changes in temperature or rainfall or other second-order environmental events. Changes in climate are decoupling the photoperiodic and seasonal signals and causing great difficulties to many species.

Over time, species have closely matched their life cycles and behaviours into an inter-related ecological web. But the difference between daytime and night-time temperatures are diminishing; animals are shifting their ranges poleward; and plants are blooming days, and in some cases weeks, earlier than they used to. Caribou in western Greenland are already suffering from what is known as trophic mismatch. The animals, close relatives to wild reindeer, are dependent on plants for all their energy and nutrients. Plants are primary producers and are known as the first trophic level. Herbivores that feed on them are the second trophic level, and then the herbivores in turn are fed on by carnivores in the third trophic level. Each trophic level has less biomass than the one below it. In the spring, caribou switch from eating lichens buried beneath the snow to grazing the new growth of willows, sedges and flowering tundra herbs. As the birth season approaches, they are cued by increasing daylength (photoperiod) to migrate inland to the western edge of the country's inland ice sheet, into areas where this newly emergent food is plentiful and where they give birth to their calves.

But these plants, which initiate growth in response to temperature rather than photoperiod, are nowadays reaching their peak nutritional value much earlier in response to rising temperatures. When the animals arrive at their calving grounds, pregnant females find that the plants on which they depend have already reached their peak productivity and have begun to decline in nutritional value (Post *et al.*, 2008). Eric Post of Pennsylvania State University, the lead investigator of the seven-year caribou study, describes what is happening when he says, 'caribou are not adjusting their birth season to help keep up with changes in plant growth. As a consequence, their food is in a sense being taken off the table again before caribou have had a chance to get what they need' (Post & Forchhammer, 2008).

What is happening to the caribou is being replicated all over the globe. Warming beyond the ceiling of temperatures reached during the Pleistocene and the change in the timing of seasonal events will stress ecosystems and their biodiversity far beyond the levels imposed by the global climatic change that occurred in the recent evolutionary past. To compound the problem, climate change and the destruction and fragmentation of habitats have confined many species to relatively small areas within their previous ranges with reduced genetic variability. It matters that we understand why the seasons happen. We can glance at the date and that tells us which season we are in. But if we have no understanding, no empathy and no sense of awe for the natural world, then we will not understand what is happening to us. *The New Yorker* magazine sent Elizabeth Kolbert round the world to write a long series of articles on climate change. She came back with many anecdotes, but the most poignant was from the Arctic. She recorded meeting an Inuit hunter named John Keogak, who lives on Banks Island, in Canada's Northwest Territories, some 800 km north of the Arctic Circle. He told her that he and his fellow-hunters had started to notice that the climate was changing in the mid-1980s. A few years ago, for the first time, people began to see robins, a bird for which the Inuit in his region have no word (Kolbert, 2005). Species are already beginning to adapt to changes in the seasons and their timing. Whether they can do it in time is a question that is more for us than for them.

# 2

# ADAPTING TO SEASONAL CHANGE

There is no season such delight can bring,
As summer, autumn, winter, and the spring.

WILLIAM BROWNE, VARIETY (1630)

It gets very cold in winter in eastern Canada, and geese and many other animals anticipate the weather and migrate. But wolves (*Canis lupus*) stay put, thanks to some remarkable adaptations that help them cope with a temperature range of approximately 50°C or more, from −30°C in the depths of winter to perhaps 20°C and higher in summer. The wolf's coat ranges from white in cold weather, to red or black in hot weather and greyish in normal weather. In the summer it sheds much of its fur so it will not get too hot and can run without getting overheated. In winter, when a wolf stands on snow at a temperature of −30°C, it has to prevent its feet from being frostbitten while also not losing heat from its body through its paws. It does this by lowering its paw temperature to close to 0°C, which is sufficient to stop frostbite, but maintains its core body temperature at 38°C. Because it can keep different parts of its body at different temperatures, Nicholas Mrosovsky, a Canadian biologist, has pointed out that this differential temperature regulation 'prevents the tissues on the extremities from freezing yet minimises the drain of heat. Local thermoregulation is supported by appropriate vascular specialisation to ensure that warm blood reaches the places it is most needed' (Mrosovsky, 1990).

Reindeer have an unusual physiological mechanism to deal with the extremes of light in polar regions; they change their eye colour and structure for summer and winter. Karl-Arne Stokkan, an Arctic biologist at Norway's University of Tromsø and the leader of a British–Norwegian team investigating reindeers in Lapland (northern Scandinavia), found that the eyes from animals in winter were deep blue whereas in the summer the eyes were yellow. This phenomenon has never been recorded before in mammals, and Stokkan points out, 'This difference suggests the reindeer alter their vision seasonally to match prevailing light conditions' (personal communication).

And just as we huddle under the bedclothes when we are cold, Arctic foxes, beavers and other animals do the same metaphorically and dig interconnecting burrows so that they can huddle together for warmth. In rather different conditions, meerkats survive the cooler, drier winter in the desert by living on a less appealing diet and spending more time in their burrows.

Animals and plants have developed very sophisticated adaptations to weather extreme environments, but these usually come at a geographical price. The fish that live in ice-cold Arctic seas die in waters whose temperature is higher than 10–20°C. Tropical fish cannot survive in water below 15–10°C. A classic case of the effect of temperature variation on geographic distribution is that of the closely related frogs in the genus *Rana*. Of the four North American species, the one with the northernmost geographical limit is also the one that is most cold tolerant, and the southernmost is least tolerant (Moore, 1939).

Plants are sessile and stuck where they are, so they have to adapt *in situ* to the changes in temperature and rainfall that come with the seasons. But plants are not helpless playthings of the environment. They can anticipate a threat before it arises. For instance, although the stomata – the pores in the leaves – open and close on a daily basis timed by a circadian rhythm, there is also a response that enables the pores to start to close down when the roots detect dry soil and long before there is any change in the amount of water reaching the leaves. The mechanism seems to be a chemical signal that the plants can send to the leaves, leading to the closure of the stomata before the plant experiences a water shortage (Zhang *et al.*, 1987). The plant's wide range of sensory apparatuses detects light, gravity, chemicals, predators, pests and vibration.

At high latitudes, the shortening days of autumn act as a seasonal signal to trigger bud dormancy and cold hardiness in plants, responses that enable them to survive the winter. Other survival strategies include the formation of storage organs such as bulbs or tubers. Plants time their germination so that the delicate shoots emerge when the worst of the weather has gone. For some plants the heat can be as threatening as the cold. Summer conditions can be unfavourable, with temperatures too high for normal survival. In some desert species, bud dormancy is induced by increasing daylengths; this is a protection against the heat. In very hot weather conditions, snails seal themselves up in their shells for a prolonged period and reduce their metabolic rate. The general term used to describe sustained adaptive responses to hot weather, usually involving a decrease in metabolic rate, is aestivation.

A much deeper understanding of how organisms adapt to seasonal change is emerging and is part of a trend in biology that considers the whole life cycle of an animal or plant. In his book *Biased Embryos and Evolution*, Wallace Arthur's message is that with regard to animals we have to shift attention from the individual at a narrow, specific point in its development and consider the egg, embryo, larva, pupa, foetus, youngster, mature adult, senescent individual. All these phases in the life cycle are affected by mutations, by developmental bias, and by natural selection (Arthur, 2004). This is meat and drink to those who work on seasonal change, because life cycles are in effect the means for organisms to survive and reproduce through such regular shifts in the environment. All that hard work collecting life-cycle data year after year in often arduous field conditions has become extremely relevant. The whole organism and its life cycle now matter; John Tyler Bonner, one of biology's 'grand old men' and still publishing at the age of 87, put it well when he wrote, more than 30 years ago, 'organisms do not have life cycles, rather they are life cycles' (Bonner, 1974).

The synchronising of life cycles to the environment may need to be so precise that a few days can mean the difference between survival and death. In particular, the timing of reproduction and growth is under strong natural selection pressure because there is often only a short period of favourable conditions in the annual cycle. The conditions include not just features such as rainfall and temperature, but also the whole ecological web. In these multi-trophic systems, the period during which these favourable conditions occur is to a large extent determined by the timing of growth or reproduction of species at the lowest level of the multi-trophic interaction. The organisms higher up the chain have to time their activities to the rise and fall in the abundance of those lower down the food chain.

The delicate seasonal timing balance of what has become an exemplar plant–insect–bird triad has been studied by Marcel Visser and his colleagues at the Netherlands Institute of Ecology. For many years they have watched the annual interplay of trees, moths and birds in the oak forests around Arnhem in Holland.

The winter moth (*Operophtera brumata*) is a pest in northern Europe and increasingly so in North America. As a green caterpillar, it munches on leaves and buds and is particularly fond of apple and cherry, but more or less any deciduous tree will do, and it is quite partial to oak leaves (*Quercus robur*). The moth itself usually emerges from the soil in late November and can be active into January. The male moths have four wings but the female is wingless and cannot fly. She gets her mate by emitting a pheromone that often attracts clouds of male moths. After mating, the female crawls up a tree and deposits an egg cluster on tree trunks, on branches, in bark crevices and under bark scales. The adult moth then dies and the eggs overwinter.

Eggs hatch in the spring just before bud burst and leaf opening of most of the host plants. The caterpillars feed voraciously for four to eight weeks and then spin a thread to lower themselves to the woodland floor, where they pupate in the litter. They stay in the soil in the pupal stage until they emerge in late November as adult moths.

Visser knows from records collected by his predecessors that the caterpillar must hatch at almost precisely the time of bud burst. If it hatches more than about five days early it may starve. More than two weeks too late and it will also starve, because oak leaves become infused with inedible tannin. Either circumstance leads to a lower weight at pupation or to a longer larval period, resulting in a higher probability of being eaten. There is also an increased risk of falling victim to parasitic flies such as *Cyzenis albicans*. The fly lays hundreds of eggs inside tree buds next to the growing winter moth caterpillars. They consume the parasite eggs while feasting on the vegetation during their development into pupae. The caterpillars never develop because they have eaten the parasite eggs. The parasite's eggs hatch and the larvae eat the caterpillar from the inside, using the host to incubate flies that go out to infect more winter moth larvae. The flies are dependent on the moths to breed, so their population declines as the moths die off.

As the date of the oak bud burst, which is cued by the combination of winter and early spring temperatures, varies from year to year, the eggs of the winter moth also need environmental cues to time their hatching and achieve synchrony with bud burst. It is the pattern of winter and spring temperatures and hence the relationship between frost-days (below 0°C) and those days with a temperature above 3.9°C rather than mean temperature alone that seems to determine egg hatching (Visser & Holleman, 2001).

Great tits (*Parus major*) are keenly tuned to the timing of the moth's life cycle. The great tit is usually a single-brood bird and only gets one annual go at breeding, so it has to make the most of it. And there is no better food for fledgling chicks of the great tit than protein-rich winter moth caterpillars. Great tits have eight or nine chicks in a brood, and each will eat about 70 caterpillars a day, which is about 90 per cent of their food intake. It takes about 18 days for the eggs of the great tit to incubate and hatch, so to make the most of the caterpillar splurge the timing becomes critical. If moths and birds both get the timing right, the caterpillars emerge at just the right time: they can guzzle on the new oak leaves and their population peaks just as the hungry chicks need feeding. If the parent tits are a bit late, or – to put it another way – if the caterpillars are early, then the chicks hatch after the caterpillar biomass has peaked and is on the wane, so that food is less abundant. In these conditions, the literal early bird really will catch the worm, or rather, caterpillar.

In the past 20 years, Arnhem spring temperatures have risen, and the result is some severe mistiming. Oak bud burst now occurs about ten days earlier than it did 20 years ago. Caterpillars hatch 15 days earlier, overcompensating by five days for the change in the oaks. The caterpillars were already hatching several days before bud burst in 1985, so now they must wait on average about eight days for food and they get very hungry. But the chronology of the tits in the Arnhem forests has not matched the changes in its prey.

Great tits are phenotypically plastic in their timing of reproduction, which means that they are able to adjust to some extent to changing conditions and tend to lay earlier in a warm spring. But their reproduction is complex and involves more than just egg laying. The birds forage predominantly within larch (*Larix decidua*) and birch (*Betula pubescens*) trees during the egg-laying period, eating insects, spiders and buds, but forage on oaks while rearing chicks. The bud burst of the larch and birch is much less dependent on temperature than that of oak, and their bud burst has not advanced over the 23-year period. So the timing of the food resources needed to produce eggs has advanced only marginally compared with that needed for chick rearing. Further, even if the birds lay their eggs at the same time as the caterpillars start developing, warmer temperatures afterwards cue the caterpillars to speed up their development, so the young birds hatch later than the peak in caterpillar biomass (Visser *et al.*, 1998).

Birds can reduce the interval between laying and hatching by laying smaller clutches, shortening the gap between the first egg and the onset of incubation, or reducing the duration of the incubation period. However, the timing of reproduction in the great tits at Arnhem has not advanced in step with early peak availability of food for the young over a 23-year period. This mistiming means that even if the animals can respond quickly, climatic change may not always act uniformly on all parts of the breeding season, and constraints and cues may not alter in step with selection pressures acting later in the breeding season.

Over thousands of years of evolution, the oak, moths and great tits had come to synchronise their life cycles using, in large part, temperature cues. But temperatures have risen since 1980 and this has had the effect of decoupling the cues so that the old rules do not work any more (Grossman, 2004). If something happens that upsets that timing, as has been happening in the oak forests of Arnhem, then linkages are broken and whole webs can collapse (Visser & Holleman, 2001).

This seems to be true around Arnhem, where the breeding time of the great tits is advancing each year while the emergence of caterpillars is advancing three times faster, but in Oxford the birds seem to be responding more rapidly. A long record of great tits' behaviour in a breeding site at Wytham Woods demonstrates that they have adjusted their behaviour and are laying eggs now about two weeks earlier than they did when records began about 50 years ago.

The change has been too fast to be attributed solely to evolutionary pressure. 'You would get a lag with evolution, because it takes a generation or more to have an effect,' says Ben Sheldon of the University of Oxford, who has been leading the study. 'It wouldn't manage to keep track so closely' (Charmantier *et al.*, 2008). It is the plasticity of the birds of Wytham Woods that has enabled them to shift their breeding times, and the outcome has been to maximise the chances of their chicks' survival.

The British birds have done well, but as temperatures continue to rise, the birds may reach a limit as to how far they can adjust through their plasticity and then natural selection could kill off many of them. Visser says (Inman, 2008):

> It's striking that the Oxford birds have enough plasticity to adapt to climate change. The difference might be because, around Oxford, both the early spring and late spring have warmed up, while in the Netherlands the early spring has not warmed much. So it looks like all the birds are following a rule – 'breed earlier in warm years' – that does not work well in the Netherlands. ... It's kind of a coincidence that the rule the Oxford birds use happens to work.

The Dutch and English groups are collaborating to try and find out which situation is the normal one and which is the exception.

Not all animals have been as successful as the great tits of Wytham Woods. The North American wood warblers (*Parulidae* sp.) have not adapted their migration patterns to the earlier emergence of caterpillars in its breeding ground (Strode, 2003), and the Dutch honey buzzard (*Pernis apivorus*) is also failing to adapt to the earlier appearance of the wasps that it eats (Visser & Both, 2005). The red admiral butterfly (*Vanessa atalanta*) is arriving on the UK's shores earlier from its winter grounds in North Africa, but the staple food of its larvae, the common nettle, continues to flower at the same time each year.

Visser's team has been studying the pied flycatcher (*Ficedula hypoleuca*), which also lays its eggs in the oak forests around Arnhem and feeds its young on winter moth caterpillars. Pied flycatchers overwinter in dry tropical forest in West Africa about 10° north of the equator and breed in temperate forests in Europe. Although temperatures at the time of arrival and the start of breeding by pied flycatchers (16 April to 15 May) increased significantly over the period 1980–2000, the birds have not advanced the spring arrival on their breeding grounds. They have, though, advanced their mean laying date by ten days (Both & Visser, 2001).

The flycatchers' spring migration timing is triggered in West Africa, some 4,500 km from Arnhem, by an internal circannual clock that is fine-tuned to daylength (Gwinner, 1989). Although the pied flycatcher can respond to any naturally occurring variation in the start of the spring by truncating the egg-laying process, the earlier and consistent onset of rising temperatures in the past two decades has left them with too little time in which to adjust. A significant part of the population is now laying too late to exploit the peak in insect abundance. This altered timing and resulting food shortage has led to a population decline of 90 per cent over the past two decades in areas where the food peaks earlier (Both *et al.*, 2004).

Other migratory birds could suffer similar declines. The decision on when to start spring migration becomes maladaptive if the cue used for migration is independent of the environmental change in the breeding area, which is the case for long-distance migrants.

The Arnhem triad of oaks, moths and birds is a simple food web. These webs can become incredibly complex, particularly in the seas. When Peter Yodzis examined one 29-species food web to determine whether culling Cape fur seals (*Arctocephalus pusillus pusillus*) would increase hake (*Merluccius* spp.) biomass, he noted that the entire interaction is so complicated that even if he counted only paths with eight links, there were 28,722,675 distinct simple open pathways (paths that never pass twice through the same species) through the food web from seals to hake (Yodzis, 2000). His counter-intuitive conclusion was that culling fur seals would probably do more harm than good to commercial fish such as hake!

Cape fur seals live in the waters off the southwestern coast of Africa. The Benguela ecosystem, as the area is known, is the site of a huge seasonal increase in phytoplankton that supports a major fishing industry. A Cape fur seal consumes approximately its own mass annually in hake. So common sense would suggest that regularly culling seals so as to reduce seal biomass by a given amount will permit an increase by the same amount in the annual yield of hake biomass to the fishery. But it does not work like that.

Although seals eat hake, they also eat species that the hake feed on, and several predators of hake as well as several competitors of hake. The hake predators and the species on which the hake feeds, as well as its competitors, are in turn predators and prey of other species. So what seems simple turns out to be fiendishly complicated as the interactions come into play.

All the food webs in the seas, such as that involving the seals and the hake, are based on phytoplankton, which form the lowest trophic level and whose biomass fluctuates with the seasons. The trillions of tiny phytoplankton – mainly unicellular algae – are the world's main photosynthesisers, producing an estimated 50 per cent of the earth's free oxygen.

Phytoplankton help give the seas their green colour. Where phytoplankton production is small, such as in the Sargasso Sea, the water is deep blue. By contrast, coastal waters rich in phytoplankton are green. The total amount of these short-lived organisms produced each year weighs more than the annual growth of all terrestrial organisms combined and forms the basis of the marine food webs, which produce about 90 per cent of the world's biomass.

The very small phytoplankton are eaten by animals that are almost as small, called zooplankton I. Slightly bigger zooplankton II eat zooplankton I, and in turn small fish such as anchovy and other animals such as crustaceans eat the zooplankton II. These in turn are eaten by bigger fish and mammals, sharks, dolphins, seals and killer whales. In the seas, all of these animals derive their nourishment and energy either directly or indirectly from the phytoplankton. When the phytoplankton population grows, as it does at the seasonal blooms, the numbers of zooplankton, fish and sea mammal also increase. If the phytoplankton population declines, so must all the creatures that depend on them. As a rule of thumb, as we move up each trophic level of the food web, the biomass declines by about 90 per cent. So there is 10 per cent as much zooplankton 1 as there is phytoplankton; 10 per cent as much zooplankton II as zooplankton I, and so on upwards. The productivity of the phytoplankton and consequently the absolute mass available sets an upper limit on the size of the entire food web.

Phytoplankton are plants and reproduce very quickly when there is plenty of sunlight and nutrients. The nutrients in the upwelling arise from 180° shifts in the direction of wind as a result of land/water heating differences between summer and winter. As land has a lower heat capacity than water, in summer it heats up considerably more than water and the heated air over land rises, thus inviting winds to blow towards land from water (onshore). In the winter, land cools more significantly than water so the relatively warm air over water rises, thus inviting wind to blow towards water from land (offshore). In the first case there is a pile-up of water at the coastline. In the second case there is a removal of water from the coastline. The wind blows across the ocean surface and pulls the surface water with it. As the surface water leaves an area, the 'hole' left behind is filled in by cold, nutrient-rich water 'upwelling' from below and coming to the surface from depths of over 50 metres. When spring turns to summer, nutrients in the surface layer are consumed by phytoplankton, reducing nutrient availability at the surface. As summer sets in, phytoplankton die and drift to the bottom, taking the nutrients they ingested with them. Surface waters are now left with few nutrients available. In cooler autumn days, a limited amount of vertical mixing brings nutrients up from below. In winter, heavy winds and plummeting temperatures cause strong mixing again. In general, when nutrients are found near the surface, they are not plentiful at deeper levels and vice versa.

Blooms can be quick events that begin and end within a few days or they may last several weeks. They can cover hundreds of square kilometres of ocean. In the Gulf of Maine, spring and autumn blooms occur on an annual basis, and smaller blooms have also been detected during other times of the year.

The Benguela upwelling usually occurs in June and, according to folklore, after the last aloe plant has bloomed. At that time of year, the cold currents of the South Atlantic shift northwards, carrying with them millions of sardines. Sardines (*Sardinops sagax*) live in water at a temperature between 14 and 20°C, feeding on phytoplankton and zooplankton. Adult sardines, which are about 18–20 cm long when two years old, aggregate and spawn on the Agulhas Banks off the southern Cape coast in the spring and summer months. The fertilised eggs drift with the current in westerly and northerly directions into the nutrient-rich upwelled waters off the west coast, which are full of plankton. Here the larvae mature and develop into juvenile fish, which, once strong enough, aggregate into dense shoals and migrate southwards, returning to the Agulhas Banks to complete their life cycle.

During the winter months of June and July, cooler water moving eastwards along the eastern Cape coast towards Port St Johns effectively expands the suitable habitat available for sardines. From Port St Johns a cool, northerly flowing counter-current flows inshore of the southerly flowing Agulhas Current, carrying large shoals of sardines farther north in what has traditionally been known as the 'sardine run' (Aitken, 2004).

The cool band of water inshore is crucial to the run. If the water is too warm (more than 20°C) the sardines will remain in the cooler water farther south or move northwards farther offshore and at greater depths. Only 2 per cent of the sardines in the southern waters where the fish usually live make the northward trip, so it is not a migration in the true sense because the sardines do not travel for feeding or breeding purposes, but there are more than enough to form huge shoals up to five kilometres long. The sardine run is essentially a passive act by the sardines: they go with the dense food flow and simply extend their geographical range for a few months.

Dolphins, sharks, seals, whales, turtles, albatross, petrels, gannets and even African penguins synchronise their behaviours to the seasonal Benguela upwelling and head for the Transkei and Kwazulu Natal coastline. The sardines become a veritable feast for predators that attack a 'bait ball' from below and above as fish, mammals and birds crash into the prey with devastating results.

The Cape gannets that plunge repeatedly into the water to grab sardines have become so attuned to this sardine run that they are thought to time their breeding cycles so that their young are fledged at the time of the event. This allows the youngsters to be exposed to an abundant source of food during a crucial learning phase of their lives, thus increasing their rates of survival (Aitken, 2004).

But the study of these seasonal actions is not easy. While we continue to learn about the amazing precision by which gannets, moths and flycatchers anticipate the right time for their seasonal activities, our understanding of the circannual mechanisms increases very slowly. Mike Menaker of the

University of Virginia has pointed out that the major difficulty in the study of circannual rhythms is a consequence of the ratio of the period length of a single circannual cycle to the length of the productive life of a scientist. A two-week (14-day) experiment for a circadian biologist would take 14 years for a circannual researcher.

It makes for a tough research regime. Roland Brandstaetter, now at the University of Birmingham, emphasised Menaker's point when he wrote, in his warm obituary of his former supervisor (Brandstaetter & Krebs, 2004):

> Eberhard Gwinner started to perform a set of unique experiments where he kept birds in constant daylength of 12 hours for many years. He regularly collected data on various behavioural and physiological parameters with clockwork accuracy. Whatever he did, wherever he travelled, his diary was determined by the dates when he had to be back at his institute to investigate gonadal size, moult, or body weight of his birds. In 1967, he was the first to describe an endogenous annual (circannual) rhythm in a small migratory songbird, the willow warbler that he had observed for three years.

Over the following decades, Gwinner demonstrated circannual rhythms of reproduction, moult and migratory restlessness in several avian species, including the European starling (*Sturnus vulgaris*). Starlings maintained in a constant light/dark cycle of 11 hours of light followed by 11 hours of darkness (known as an LD 11:11 cycle) for more than four years showed a regular annual change in testis size, although in the case of this cycle it was not a 12-month but closer to a 10-month period (Gwinner, 1986). In stonechats (*Saxicola torquata*), he showed a persistent circannual reproductive rhythm over more than 10 years (Gwinner, 1996b). A core issue to him was to identify the way in which endogenous rhythms and photoperiod interact to enable correct seasonal timing, and a key aim was to achieve the sort of understanding of circannual rhythms that we have of circadian rhythms.

All multicellular organisms, and also some unicellular ones, possess a circadian (daily) clock whose basic characteristics are the same across all taxa. This alone strongly suggests that the ability to tell the time and predict change is highly advantageous to a wide variety of organisms. Organisms need to synchronise their activities to the world around them and also to control the timing of their internal processes so that they happen in the appropriate sequential fashion. Although the actual molecular mechanism of the circadian clock differs in detail across the natural world, suggesting that the circadian clock has arisen several times during the evolution of animals, plants, fungi and cyanobacteria, the principles are the same (Foster & Kreitzman, 2004).

A circadian clock effectively enables an organism to know the time of day. If the common cockroach, *Periplaneta*, for example, is put in a cockroach-sized running wheel, provided with food and water and left alone in an environmentally controlled cabinet, with the light on for 12 hours and then off for 12 hours (LD 12:12), most of its activity each day takes place during the first two to three hours of darkness. It could be that the insect is simply responding to light cues, so that when the lights are switched off after 12 hours, the insect senses the change in light levels and is immediately active for a period and then goes back to being quiescent. As the late John Brady wrote, 'this is about as unexciting a finding as banging the side of the cage every 24 hours and frightening it into activity' (Brady, 1979). If the lights are left off and the cockroach is then kept in constant darkness, there is a bunching of activity into a two- to three-hour period that recurs roughly every 24 hours. This pattern repeats itself day after day after day. Even in the constant conditions of darkness, the cockroach is able to divide time into a subjective 'day' and 'night', and anticipate what in the wild would be nighttime. In dry, scientific language, when the cockroach was placed in an unchanging or aperiodic environment (constant temperature and illumination) its daily periodicity persisted indefinitely in the absence of external driving factors. This is exciting. Clearly, there is an internal clock that ticks away at close to 24 hours even when kept in unvarying conditions, such as constant darkness.

It is not quite a 24 hour cycle, more like 24.5 hours, in the cockroach, so in total darkness the cockroach would start its subjective day a half-hour later each day. This drift is similar to a grandfather clock that runs a bit slow or fast and needs adjusting. Released from the obvious light/dark cycle that mimics the daily sunrise and sunset, the animal 'free-runs' with a natural period slightly longer than the solar cycle. The free-running rhythm is synchronised each day and forced (entrained) by light exactly to the 24-hour solar cycle. For most organisms, light is the main time-giver (*Zeitgeber* in German) that keeps the mechanism synchronised to dawn and dusk.

The cockroach clock displays all the key characteristics of all circadian rhythms. They show a free-running rhythm under constant conditions with a period that is close to, but not exactly, 24 hours; the free-running rhythm can be entrained to exactly 24 hours by an environmental time-giver; and the rhythm, like all good clocks, is temperature-compensated (Foster & Kreitzman, 2004). Likewise, the endogenous circannual clock shows similar characteristics. Under constant conditions, it has a period that is close to a year and will free-run. It is entrained by an environmental time-giver such as the daylength (photoperiod) and it seems in most cases to be temperature-compensated.

But there, to our knowledge, the similarities end. The search for the circadian clock mechanism has been difficult, but we now have a good idea of where that clock is, what it is made of and how it works. We have a broad understanding of the molecular processes that enable a near-24-hour rhythm in the rise and fall in abundance of various proteins to be sustained and entrained to the daily solar cycle by the light signal of dawn and dusk. We are some way from the full story but at least there is a proposed mechanism. The same still cannot be said for the circannual timer discovered by Ebo Gwinner and independently by the Canadian biologist Ted Pengelley. However, we have uncovered some of the mechanisms involved. How the plants and animals know the time of year is the subject of the next four chapters.

# 3

# ANTICIPATING SEASONAL
# CHANGE IN PLANTS

A time to plant and a time to uproot what is planted

ECCLESIASTES 3:1–8

The first draft of this chapter was written in early 2007 when the British newspapers and TV were full of stories describing how the change in climate was throwing plants and animals into total confusion. In northern England, the earliest spring flowers such as snowdrops and hazel-catkins had already been and gone and late spring bloomers such as gorse were flowering unusually early.

Across the northern hemisphere, lilac and honeysuckle are blooming about a week earlier than they did half a century ago. In Vermont, maple trees were historically tapped for their syrup between the middle of March and the middle of April. With warmer late winter and early spring temperatures, this has shifted towards the middle of February (Banks, 2006).

In one huge study, researchers analysed 125,000 records and observations in Europe compiled between 1971 and 2000. They looked at 542 plant species in 21 European countries. Over the three decades, 78 per cent of all leafing, flowering and fruiting records advanced (30 per cent significantly) and only 3 per cent were significantly delayed. The average advance of spring and summer was 2.5 days per decade in Europe (Menzel *et al.*, 2006). The

British father-and-son team of Richard and Alistair Fitter found that in the area around Oxford, 16 per cent of species flowered significantly earlier in the 1990s than previously, with an average advancement of 15 days in a decade (Fitter & Fitter, 2002).

As farmers, gardeners and horticulturalists have known since time immemorial, if the plants get the timing wrong then their chances of successful reproduction are slim. A very cold spell, as we had in the UK in late March 2007, threatened those plants that had come into blossom weeks in advance, as well as the bees, butterflies and those animals that had arisen from their stupor as the temperature rose.

Because plants are sessile, they cannot escape or hide from annual seasonal change by migrating or burrowing. They have to adapt their life cycle, structures and physiology in such a way that they are best able to survive and reproduce throughout the daily and annual cycles. In particular, they have to rely on synchronising their developmental stages with the seasons and effectively 'know' when to flower, when to germinate, when to produce seed and when to enter dormancy.

In temperate climes, responding to a variable environmental cue such as moisture or temperature alone as a means of coordinating with seasonal change has its dangers. A warm January or February can be followed by a cold snap in March. It is easy for a plant to be 'fooled' by a short run of higher temperatures. Although for some species there may be selective advantages in being the first to flower, because there is less competition around, the risk can be high.

The timing of the change in seasonal conditions has been predictably consistent up to now and many plants use what is effectively a calendar to 'know' the time of year and so when to do what. For these plants there is a species-specific critical daylength. This may be as short as 6.5 hours in some marigolds (*Calendula officinalis*) or as much as 16 hours for Japanese morning glory (*Ipomoea nil*). The marigolds will only flower when the days are longer than 6.5 hours, whereas the Japanese morning glory will only flower when the daylength is shorter than 16 hours.

By anticipating seasonal change correctly, deciduous trees can time their leafbud burst so that the risk from a late frost is low, but not so long into summer that their annual period of photosynthetic activity is much diminished. However, to be of such practical use, these calendars have to be synchronised to reliable natural markers of daily and seasonal change. How plants do this has been a work in progress for over two centuries and began with a study of the daily or circadian rhythms. Ubiquitous internal circadian clocks enable living things to 'know' the time of day so as to anticipate the daily fluctuations in environmental variables such as temperature, humidity, rainfall, ultraviolet flux, wind, and the intensity, spectral quality and duration of visible light.

The first person to begin a scientific investigation of this phenomenon was a French astronomer Jean Jacques d'Ortous de Mairan. He was interested in the rotation of the earth and was curious to find out why the leaves of plants were rigid during the day and drooped at night in time with the pattern of light and dark that resulted from this rotation. In 1729 he put a mimosa plant (*Mimosa pudica*) in a cupboard and peeked in at various times. Although the plant was permanently in the dark, its leaves still opened and closed rhythmically – it was as though the plant had its own representation of day and night. The plant's leaves still drooped during its subjective night and stiffened up during its subjective day (de Mairan, 1729). De Mairan had unknowingly identified the first circadian rhythm generated by an internal molecular clock. The fundamental point about these internal clock mechanisms is that they are endogenous and tick metaphorically with an inbuilt periodicity of close to 24 hours. Through a complicated process of feedback loops between genes and their protein products, a set of 'clock' genes effectively produce a rhythm that is close to, but not quite, 24 hours. The circadian clock in plants controls many regular daily processes, including leaf and petal movements, the opening and closing of stomatal pores, the production and release of pollinator-attracting fragrances, and a range of metabolic activities, especially those associated with photosynthesis.

Left to themselves in conditions of constant light or constant dark, the circadian rhythms would start to drift or free-run through time. But in the natural world the internal clock is continually synchronised to the daily solar cycle by the light signals of dawn and dusk that result from the daily rotation of the earth. Think about resetting a watch to the radio pips and you have a good analogy. The outcome is that plants and animals can predict daily and – as we will see – annual events and behave accordingly.

Although a light-entrained clock influences seasonal cycles, including flowering in plants such as camellias (*Camellia* spp.), which flower early in the year, and those such as chrysanthemums (*Chrysanthemum* spp.), which flower later, others seem to flower regardless of the daylength. For instance, temperature in spring is the main driver for flower-bud development and flowering in the common purple lilac (*Syringa vulgaris*). The triggering factors for dormancy and reaction to spring temperature are located in the branch tips (buds). Watch a lilac branch growing close to a warm, south-facing wall: spring flowering on this branch will be much earlier than on branches of the same plant farther away from the wall. T. S. Eliot referred to lilacs in *The Waste Land*:

April is the cruellest month, breeding
Lilacs out of the dead land, mixing
Memory and desire, stirring
Dull roots with spring rain.

which has led Harriet McWatters of Oxford University to ask, 'Will future annotations of *The Waste Land* have to take account of climate change?' (Personal communication).

There is a considerable selective advantage if different plant species reproduce at different times. This 'complementarity' allows many species to coexist because it reduces overlap in the time period when species compete for limited resources such as space, light and soil nutrients. So grasses tend to flower early in the growing season and wildflowers later. Plants of a particular species in the same location tend to flower at the same time every year even though they may have started growing at different times. This synchronised flowering helps promote cross-pollination.

Orchestrating this complex daily and seasonal temporal environment within individual plants and both within and between species needs a reliable marker. Because daylength follows a predictable pattern year on year, it is a clear signal for such seasonal events. The photoperiodic response tracks the expanding and contracting day or night length to mark the passage of the seasons. But it has been a long scientific slog to arrive at this seemingly straightforward understanding, and the details are still being worked out.

Carbon Dioxide ..... *Chlorophyll* ..... Oxygen

$$6CO_2 + 6H_2O + \text{light} \longrightarrow C_6H_{12}O_6 + 6O_2$$

Water ..... Glucose

**Figure 3.1** Photosynthesis is the process by which plants, some bacteria and some protistans use the energy from sunlight to produce sugar, which cellular respiration converts into ATP, the 'fuel' used by all living things. Photosynthesis is a complex reaction, involving both light-dependent and light-independent (dark) stages. The conversion of sunlight energy into usable chemical energy is driven by chlorophyll photopigments.

For the best part of 200 years, the key findings in botany were concerned with how plants grew. This focus led to an understanding that plants grow through photosynthesis, which enables them to capture energy from the sun and use it to convert carbon dioxide and water into carbohydrate (glucose) and oxygen (Figure 3.1).

In the early years of the last century, the US government put large sums of money into agricultural research because there was intense pressure to increase agricultural productivity in the USA as a result of the massive growth in population and the need both to increase output and to reduce agricultural costs. The hope was that productivity could be increased by growing bigger and better plants, and in the early part of the twentieth century tobacco farmers in Maryland became excited by the appearance of a new mutant strain. These giant tobacco plants grew up to 4.5 metres tall and put out nearly 100 leaves, more than double the height and double the number of leaves of the common tobacco (*Nicotiana alata*). This was productivity on stilts. But the appropriately named 'Maryland Mammoth' had a nasty trick up its stalk. It did not know when to stop making leaves and start making flowers and seeds and carried on growing right through the summer and autumn until the frost killed it. The Maryland Mammoth rarely flowered in Maryland and it never produced seed in the field that could be used to plant the next crop. The plant did not seem to know when to do what!

The only way to propagate the plant was to transplant the stalk to a greenhouse in the autumn and wait for it to seed. Even starving the plant to inhibit growth and promote blossoming did not work. It was not nutrition that was the problem, but the timing of development. And it was not just tobacco plants that were proving difficult. Soya beans (*Glycine max*) were not grown in Minnesota because by the time the plants flowered, the cold weather was setting in and killed them off. Farmers in Maryland could grow the beans, but they ran into difficulties when they tried to stagger the harvest time so that they could organise their work schedules and use their labour and machinery more effectively. They tried planting out the crop at two-week intervals, but irrespective of when they were planted, the plants would all set flowers and need harvesting at more or less the same time.

Both the Maryland Mammoth and the soya-bean problem were handed to Henry Ardell Allard and Wightman Wells Garner, two plant physiologists working for the US Department of Agriculture at the Arlington Research Farm in Virginia. Garner and Allard discussed the problems over a daily lunch in the staff canteen, sited where the Pentagon is now (Sage, 1992). The scientific literature was sparse. They knew that in the mid 1800s it had been suggested that daylength might influence flowering, and by the early 1900s it was known that the intensity of light, its wavelength and the duration of exposure all had to be taken into account. Julien Tournois, a Frenchman who was killed early in the First World War, had noticed in 1910 that hemp (*Cannabis sativa*) and hop (*Humulus* spp.) plants flowered very early in his winter greenhouse (Tournois, 1912). He had grasped the point that flowering occurred in these plants as a result of the duration of the daylight and even concluded that in reality it was the long nights rather than the short days that were responsible.

But the literature was still in the realm of anecdotal observation, and what was needed was controlled experimentation to test the hypotheses. Garner and Allard attacked the problem by growing Maryland Mammoth tobacco plants in pots and soya beans in troughs outdoors, and at different times of the day they moved the plants into a glorified shed that had a tight-fitting door and provided a dark chamber. The following morning they would put the plants back outside alongside controls and continue until the plants blossomed. By effectively shortening the daylength, they caused the tobacco to flower three months early and the soya beans five weeks early (Garner & Allard, 1920). It was a simple experiment, but essentially they laid the foundations for the modern, multi-billion-dollar horticultural industry, as well as leading to huge improvements in agriculture.

They labelled plants such as soya bean, which flower in the autumn, as

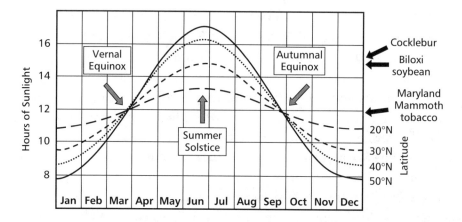

**Figure 3.2** The daylength throughout the year varies with latitude. Shown are the daylength changes for 20–50° N. For reference, Winnipeg is at 50° N, Chicago at 40° N, San Francisco at 37° N and Miami at 26° N. The required daylength to initiate flowering in several short-day plant species (Maryland Mammoth tobacco, Biloxi soybean and cocklebur) are shown at the right (arrow). Because the Maryland Mammoth tobacco plant requires a photoperiod of less than 12 hours to flower, it will flower only in the autumn. This restricts its range to relatively low latitudes because flowering at higher latitudes in the autumn would mean that the low temperature would not allow the plant to survive to seed.

short-day plants (SDPs). These flower when the daylength is shortening and has contracted to a species-specific critical length. Other plants such as barley (*Hordeum vulgare*) flower in the spring or summer when the days are lengthening, and they named these long-day plants (LDPs). Still others are not sensitive to the photoperiod and are called day-neutral plants. The Maryland Mammoth requires a daylength of 12 hours or less if it is to flower. At 40° N, the latitude at Beltsville, Maryland, to which the Arlington Research facility relocated, tobacco seeds are normally planted out in the field during April and May when the daylength is approximately 14 hours. After 21 June, the daylength begins to decrease, falling to 12 hours by 21 September, when flowering is initiated in the Maryland Mammoth – but by then it is getting too cold for the plant to survive through to seed (Figure 3.2).

As well as tobacco plants and soya beans, Allard and Garner experimented with carrots, lettuce, hibiscus, violets, goldenrod and a wide range of other plants. Their conclusion was simple but the implications were far-reaching: 'both the rate and extent of the growth attained by the plants under study and the time required for reaching and completing the flowering and fruiting stages are profoundly affected by the length of the daily exposure to sunlight' (Garner & Allard, 1920). They had firmly established the fundamental principle that the length of day (and, inversely, the night) controls the timing of flowering in many plants and named it photoperiodism. Crucially, their work showed that plants use light not just for energy as in photosynthesis but also to provide information that locates them in time. This was the key insight into the mechanism by which plants know the time of day and the time of year by synchronising their internal clocks to the external light environment.

The worldwide distribution of many plants is closely linked to photoperiod. For example, ragweed (*Ambrosia* spp.) a short-day plant, is not found in northern US states because the plant flowers only when the daylength is shorter than 14.5 hours. This does not happen in northern Maine until August. This is so late in the growing season that the first frost arrives before the resulting seeds are mature enough to resist the low temperatures, and so the species cannot survive there. By contrast, spinach (*Spinacia* spp.) a long-day plant, is not found in the tropics because there the days are never long enough to stimulate the flowering process. Maryland Mammoth and Biloxi soya beans, as well as chrysanthemums and others, flower in reply to the shortening days of late summer or early autumn. Lettuce (*Lactuca* spp.), spinach and the like are long-day early-summer flowers. Cultivated strawberries can be short-day or long-day plants depending on the variety. 'Junebearing' strawberries that bloom in response to short daylengths are planted in late autumn or winter and produce fruit the spring after planting. 'Everbearing' strawberry varieties bloom under long-day conditions and fruit two or more times per season. And day-neutral plants such as tomatoes (*Lycopersicon* spp.) and dandelions (*Taraxacum* spp.) use mainly temperature as the seasonal cue and can flower right up until frost. Photoperiodic plants that require a specific daylength threshold to induce the transition from vegetative to reproductive growth are called obligate plants, whereas facultative plants will flower eventually even if the critical length of day is not reached.

A large number of plant activities are regulated by photoperiod, including: the development of reproductive structures in mosses and in flowering plants; the rate of flower and fruit development; stem elongation in many herbaceous species as well as coniferous and deciduous trees; autumn leaf drop and the formation of winter dormant buds; the development of frost hardiness; the formation of roots on cuttings; the formation of many underground storage organs such as bulbs (onions (*Allium* spp.)), tubers (potato (*Solanum* spp.)) and storage roots (radish (*Raphanus* spp.)); runner development (strawberry (*Fragaria* spp.)); the balance of male to female flowers or flower parts (especially in cucumbers (*Cucumis* spp.)); the ageing of leaves and other plant parts; and even obscure responses such as the quality and quantity of essential oils produced by jasmine (*Jasminum* spp.).

Since Garner and Allard's seminal discovery, thousands of papers have been published in the long attempt to understand how plants manage to measure relative daylength and use the signal to time their development. The key problems were how light was detected, how daylength was measured and how flowering was initiated.

Much of the work in the 1930s went into the last of these questions. The Russian scientist Mikhael Chailakhyan proposed that there was a chemical messenger, which he named 'florigen', that was the signal to flower. In 1938, working with the short-day plant cocklebur (*Xanthium* spp.), Karl Hamner, another Beltsville alumnus, and James Bonner demonstrated that mature leaves are crucial for reception of the daylength signal. Defoliated plants never flowered; Hamner and Bonner cut away leaves to see how much was needed for flowering to take place. Just two to three square centimetres was enough. The leaves received the light signal, but because flowering takes place at the tip of the plant there had to be a means within the plant for communicating the information. They showed by grafting experiments that the signal travelled not only within the plant but also from one plant to another, so it had to be a diffusible substance. Mikhael Chailakhyan was right, and some 70 years later and after many research efforts, a team led by George Coupland at the Max Planck Institute for Plant Breeding Research Department in Cologne is unravelling the molecular processes involved (Corbesier & Coupland, 2006).

But Hamner and Bonner's key finding in 1938 was that in the cocklebur, flowering 'is not primarily a response to duration of the photoperiod, but rather a response to duration of the dark period' (Hamner & Bonner, 1938).

Short-day plants were really long-night plants. Cockleburs initiated flowering when they had been kept in the dark for at least 8.5 hours, and it was this dark period that was critical.

Their experimental procedure that led to this finding was satisfyingly elegant and simple (Figure 3.3). The particular species they used, *Xanthium strumarium*, initiates flowering when there is less than 15.5 hours of daylight. They reasoned that if the length of daylight is the determining factor then the plants should flower on any cycle having a daylength of less than 15 hours and fail to flower if the daylength is more than 16 hours. If the dark period is critical, then the plants will fail to flower on any cycle having a dark period of less than 8.5 hours, irrespective of the length of the photoperiod. *Xanthium* flowers profusely when it is kept in a 24-hour cycle of 16 hours of darkness and 8 hours of light. So in one experiment, plants experienced a 12-hour cycle, of which four hours were illuminated and eight were dark. Other plants went through a 48-hour cycle, consisting of 16 hours in the light and 32 hours in darkness. The plants on the 12-hour cycle remained vegetative, whereas those on the longer cycle, which included both a long dark period and a long photoperiod, flowered.

To clinch the argument, Hamner and Bonner gave some plants a nine-hour dark period, but broke it into two 4.5-hour dark periods with one minute of exposure to artificial light. This break was enough to prevent the initiation of flowering (Figure 3.3). When the converse was done with other plants, interruption of the light period with a dark pulse had no effect. They concluded that flowering could be initiated by just one dark period of the critical length; this meant that the flowering process could begin on a specific day of the year, but only if the temperature was about 21°C or more. At lower temperatures, several successive dark periods above the critical length were required. The temperature effect varies with species. Poinsettia (*Euphorbia pulcherrima*) and morning glory (*Ipomoea purpurea*) are short-day plants at high temperatures and long-day plants at low temperatures.

Horticulturists leapt on this finding as they quickly realised that instead of leaving the lights on for several hours in their greenhouses each night to create a short night so as to accelerate flowering, they could save a fortune on light bills by using a few minutes of light to get the same result. In the late 1930s, Harry Borthwick, who was working at Beltsville at the time, decided

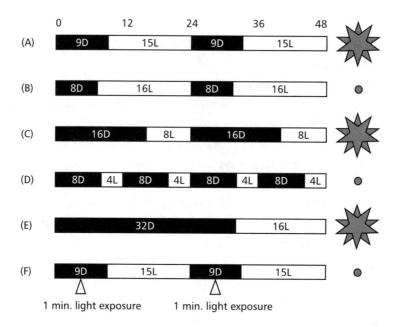

**Figure 3.3** *Xanthium strumarium* was used by Hamner and Bonner to determine whether it is the duration of the light or dark period that dictates flowering. They reasoned that if the length of daylight is the determining factor, plants should flower on any cycle having a light duration of less than 15 hours and should fail to flower if the length of the light period is more than 16 hours. In contrast, if the dark period is critical, the plants will not flower on any cycle having a dark period of less than 8.5 hours, irrespective of the length of the photoperiod. (A) A photoperiod of LD 15 : 9 triggers flowering; (B) a photoperiod of LD 16: 8 fails to induce flowering; (C) LD 8 : 6 induces flowering; (D) LD 8 : 4 fails to induce flowering; (E) LD 16 : 32 induces flowering; (F) 9D (interrupted by one minute of light):15L fails to induce flowering. Collectively these data led Hamner and Bonner to conclude that flowering could be initiated by a dark period longer than about 8.5 hours.

that rather than work like so many others on the initiation of flowering itself, a more fruitful approach would be to start from the other end and investigate the reception of the light signal.

Borthwick focused on the action of light in the photoperiodic process and he began by measuring the action spectrum of plants. This was based on the simple idea used by Tomas Engelmann to measure the first action spectra of living cells in the late 1800s. Engelmann projected a spectrum of light on *Cladophora*, an aquatic algal filament full of chlorophyll-containing chloroplasts. He noted through his microscope that oxygen-seeking

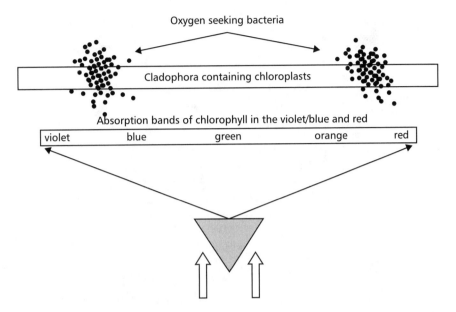

**Figure 3.4** Engelmann projected a spectrum of light onto the chloroplasts of a filamentous green alga and observed that oxygen-seeking bacteria introduced into the system collected in the region of the spectrum where chlorophyll pigments absorb.

bacteria introduced into the system collected in the blue and red regions of the spectrum, the regions where chlorophyll pigments absorb light most strongly and thus produced oxygen (Hangarter & Gest, 2004). Chlorophyll reflects light in the green band, which is why much of nature looks green to us.

An action spectrum is a graph plotting a biological event on the *y*-axis against the wavelength of the light used on the *x*-axis. An absorption spectrum, on the other hand, is a measure of the amount of light absorbed by a chemical on the *y*-axis as a function of the wavelength of light applied (*x*-axis). If the shapes of an absorption spectrum and an action spectrum match, one can deduce that it is the chemical that is involved in the light-absorbing process that leads to the biological activity (Figure 3.5).

Borthwick soon found that obtaining an action spectrum of whole plants was not easy. The light sources had to be big enough and bright enough to provide a uniform radiation field over a large area; they had to be of a sufficient intensity to produce a physiological action in short periods of

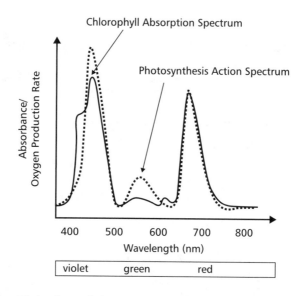

**Figure 3.5** Simplified action and absorption spectra of chlorophyll. The close match between the action spectrum and the absorption spectrum allows the biochemistry of the light-responsive pigment to be deduced.

time; and there had to be a high spectral purity so that the light was mono-chromatic at a given wavelength.

He asked Sterling Hendricks for help. Hendricks was already something of a legend. He had trained as a chemist and was Linus Pauling's first graduate student. He was also a world-class mountaineer. Using a large lamp that had somehow been 'liberated' from a Chicago burlesque theatre and two large glass prisms that had first seen active scientific service during Victorian times, their experimental set-up enabled them to pass light through the prisms and cast a spectrum 16 metres away in a 2.2-metre swathe (De Quattro, 1991).

They quickly discovered that red light in the region of 660 nanometres (nm) was the most effective in inducing seasonal effects in plants. Their breakthrough came when they re-discovered an earlier finding that whereas lettuce seeds maintained in the dark have low rates of germination, when exposed to white light they have high rates of germination. More precisely, red light at 660 nm will induce the lettuce seeds to germinate, but far-red light at 730 nm inhibits lettuce seed germination (Borthwick *et al.*, 1952).

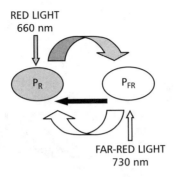

RED LIGHT
660 nm

$P_R$

$P_{FR}$

FAR-RED LIGHT
730 nm

**Figure 3.6** Phytochromes exist in two interconvertible forms. $P_R$ was so named because it absorbs red (R; 660 nm) light, and $P_{FR}$ because it absorbs far-red (FR; 730 nm) light. Absorption of red light by $P_R$ converts it into $P_{FR}$. Absorption of far-red light by $P_{FR}$ converts it into $P_R$. In the dark, $P_{FR}$ spontaneously converts back to $P_R$.

The promotion–inhibition was repeatedly reversible (they got as far as 100 cycles) and this led them to a simple explanation of the effect: a reversible, light-sensitive pigment existed in two interconvertible forms, one $(P_R)$ absorbing red light maximally at 660 nm, and the other $(P_{FR})$ absorbing far red light maximally at 730 nm. The two states could be switched from one form to the other by light. Furthermore, in the dark, $P_{FR}$ spontaneously converts back to $P_R$ (Figure 3.6). This pigment was thought to regulate the metabolic pathway that led to germination and other light-sensitive physiological responses such as the initiation of flowering. In essence, they thought the pigment, which they dubbed phytochrome, acted as a light-sensitive switch.

The reversible phytochrome process was a possible timing mechanism that has been likened to the action of an hourglass. In an hourglass, sand runs from one half into the other, and at any moment the relative amount of sand in the bottom and the top is an indication of how many hours have passed (Figure 3.7). The idea was that, at the beginning of the night, molecules of $P_{FR}$ begin converting to $P_R$; this continues until daylight, when the process reverses. At any point in the night, the amount of $P_{FR}$ is proportional to the duration of darkness at that point. When $P_{FR}$ is at a critical level during the night, the flowering reaction would be initiated or inhibited. So the plant, using this hourglass timer, could measure night length and therefore calibrate its behaviour to the time of year. In the case of a long-day

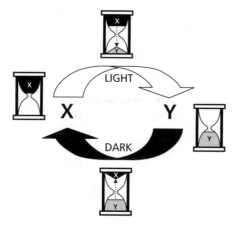

**Figure 3.7** The 'hourglass' model of photoperiodic timing. In the light, substance X is converted to substance Y. In the dark, substance Y is converted back to substance X. The duration of the light and dark periods will determine the ratio of X to Y. Such a ratio could be used to measure day or night length and hence the season.

plant, the night would be relatively short and, with less time for the $P_{FR}$ to reconvert to $P_R$, flowering would be initiated at the critical level of $P_{FR}$. But in a short-day plant there is a longer period of darkness and so there would be more time to convert the $P_{FR}$ to $P_R$. The short-day plant would flower at the lower critical level of $P_{FR}$. To 'reset' an hourglass, you simply turn it over. The flowering timer would be reset during the day, when $P_{FR}$ levels are re-established.

It was a potential explanation to a biological problem – what mechanism could possibly produce a timing signal that was sensitive and synchronous to the external environment? In the 1950s there was none of the understanding of the molecular processes that we have now, but elegant and simple as the hourglass model was, it was wrong! When levels of $P_R$ and $P_{FR}$ were actually measured and the time of conversion was determined, it turned out that in some species the bulk of $P_{FR}$ to $P_R$ dark conversion took less than four hours, far too low to account for critical night lengths of 12 hours or more. This was only one of several profound flaws.

All the time that the Beltsville team was working on the hourglass model, an alternative explanation for the mechanism of biological clocks that had been formulated in the mid-1930s was lying unnoticed in a German periodical. Erwin Bünning was the scientist who, working more or less on his own and with an apparatus budget of $25 a year, laid the foundations for the rigorous analysis of biological rhythms. In a remarkable series of experiments, Bünning, while still in his twenties, established the fundamental principles of circadian rhythms and in an outstandingly prescient study he provided a convincing explanation for the seasonal rhythms in plants. Had he not been an unwilling soldier in the German army in the Second World War, he might well have received even higher recognition for his undoubted genius.

Bünning began his pioneering studies of biological rhythms in 1930 when he found that the leaves of the common runner bean (*Phaseolus*) are elevated during the day and lowered at night. He measured the changes in leaf position by 'wiring' them up to a lever that would trace leaf movement on a drum that rotated very slowly (a 'smoke drum'). Bünning showed that this rhythm persists when plants are kept in constant light and that this oscillation has an average period of 24.4 hours (Bünning, 1973). This drifting pattern of an organism kept under constant conditions of darkness or light is known as its 'free-running' rhythm (Figure 3.8). These near-24-hour rhythms are endogenous. A light/dark cycle, or other 24-hour cues, can synchronise the rhythms but they do not cause them. It is hard to overstate the conceptual significance of Bünning's work. Over the centuries there had been many observations that plants and animals carry out their activities with regularly timed rhythms, 'but these were no more than interesting facts of nature until the emergence of the concept of the clock' (Ward, 1971).

When Bünning published his hypothesis that organisms possess endogenous clocks, and for decades afterwards, there was no idea as to the molecular mechanism that could produce such a rhythm. We know now broadly how it works, and David Krakauer of the Santa Fe Institute summarises it well when he says (Krakauer, 2004):

> The basis of the organismal circadian rhythm is an elaborate gene regulatory network within cells that produces cells capable of synthesising proteins whose concentrations vary in a periodic fashion. This periodicity results from a combination of negative feedback and time delays. It transpires that organisms are aggregates of thousands of cellular clocks, each beating out rhythms according to the statistical

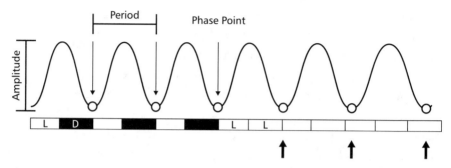

**Figure 3.8** Movement of a plant leaf under a 24-hour light (L):dark (D) cycle and then under constant light. The magnitude of leaf movement is the amplitude of the rhythm. The phase point is any defined point in the cycle. In the diagram the phase has been defined as the low point of the oscillation in leaf movement. The period of the rhythm in leaf movement under a 24-hour L:D cycle is 24 hours. Note that the phase points occur at the same time. Bünning showed that under conditions of constant light these rhythms persist but 'free-run' with a period that deviates from 24 hours. In the figure the phase points get later (thick arrows), showing that the endogenous clock is longer than 24 hours. In nature, 'dawn' and 'dusk' act as zeitgebers and entrain circadian rhythms to 24 hours; in the laboratory the same entrainment occurs when the lights are switched on and off.

regularities emerging from their own feedback loops, and coming together as a statistical unity to give us our sense of time.

The basis of the clock is a rhythm generated by a feedback loop, which is shown in its simplest form in Figure 3.9.

Multiple 'clock' genes and their proteins participate in this 24-hour loop, and receptors sense light and align this molecular oscillation to the environmental light/dark cycle. Output molecules then convey the temporal information into physiological responses. Although most biochemical reactions take mere moments to complete, the time-delayed periodic synthesis and degradation of proteins produces a near-24-hour rhythm. The regular, periodic production of proteins resulting from a combination of negative feedback and time delays is now the conceptual basis of virtually all our modern biological understanding of the generation of rhythmic patterns at the molecular level.

The key to peering into the black box of the clock and unravelling the molecular mechanisms was provided by Crick and Watson's 1953 discovery

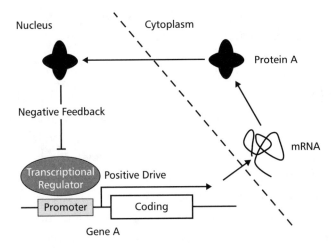

**Figure 3.9** A simple feedback loop. A transcriptional regulator protein binds to the promoter region of 'gene A' (DNA) and activates transcription from the coding region of the gene. After translation of the messenger RNA (mRNA) into 'protein A' in the cytoplasm, protein A moves into the nucleus and blocks the transcriptional drive of its own gene. Lack of transcriptional drive blocks further mRNA production and hence the production of protein A. Degradation of protein A in the nucleus allows, once again, the transcriptional regulator to produce more mRNA and then protein A. Protein A moves into the nucleus and blocks transcription – and the cycle continues. Note: transcription is the process by which a gene's DNA is encoded into a strand of mRNA. Translation is the process by which the information encoded in the mRNA is translated into the sequence of amino acids that constitute a protein.

of the structure of DNA and the subsequent leaps in understanding provided by molecular biologists. But credit also has to go to a weed that enabled scientists to study in detail the implications of Bünning's circadian and seasonal work. Research in molecular biology needs cheap and available model organisms that reproduce quickly. It took a while before the botanists found a plant for their researches that proved to be as important to them as the fruit-fly and the mouse have been for animal genetics.

Thale cress (*Arabidopsis thaliana*) is as insignificant a plant as they come. It is a member of the mustard family, and many of the tiny plants can be grown in a small area. It has a short life cycle of six to eight weeks, produces several thousand seeds per plant and has only five chromosomes. The plant is amenable to classic genetic manipulation by self-fertilisation or crossing. The genome has been fully described, and *Arabidopsis* is a small, fast-breeding plant that is almost perfect for detailed genetic study. The only potential drawback is that it is a long-day plant and so its regulation of flowering time is most probably not identical to that in other species, in particular short-day plants.

*Arabidopsis* is the favoured species for study not only because it is small and grows fast, but also because a range of mutations can alter various aspects of the photoperiodic control of flowering. For example, mutations in certain genes convert these long-day plants into day-neutral plants that can flower rapidly in short days, whereas mutations in other genes result in day-neutral plants that exhibit a long delay in flowering.

Although *Arabidopsis* clock and light signalling mutants can be picked out by obvious physical characteristics such as height (hypocotyl length), it was hard to find a technique that could enable rhythms to be monitored easily. The leaves of an individual plant can be 'wired up' as Bünning did in the 1930s, but it is difficult to automate and impracticable in large numbers.

Steve Kay and Andrew Millar genetically engineered *Arabidopsis* plants so that firefly luciferase serves as a marker or 'reporter gene' for the clock. Now, when researchers want to study *Arabidopsis* rhythms, they spray some of the specially prepared tiny plants with a fine mist of luciferin, the small organic molecule that gives the firefly its fire. The plants begin to glimmer a few hours before dawn, growing steadily brighter through the morning and weaker as the day wears on. This elegant way of giving the plant's clocks a convenient pair of hands is a lot easier than monitoring leaf movements for days on end, and turned *Arabidopsis* into the botanical world's mouse for the purposes of genetic investigation of the molecular clockwork. It also enabled plants to be monitored in the dark with closed-circuit television.

Kay, Millar and their colleagues have come up with a description of what actually goes on in the plant circadian clock (Yanovsky & Kay, 2003). It is best understood with the aid of the diagram in Figure 3.10, which by itself does less than justice to years of hard and imaginative work. This diagram, together with the following figures, describes in a simple way how the circadian rhythms interact with daylength to provide a seasonal marker.

The key insight is based on Allard and Garner's observation that although plants need light to provide them with the means of fixing the carbon in the air to provide themselves with a store of energy, they need it also for information. Plants can sense, evaluate and respond to light quality, quantity, direction and duration with a highly sophisticated suite of photoreceptor pigments. The ability to sense the environment is so important that in *Arabidopsis* about 25 per cent of its 25,000 genes are involved in signalling reception and communication, gathered under the term signal transduction (Michael and McClung, 2003).

The light signal is received and transduced by three different sets of photoreceptive pigments. There are five versions of the phytochrome pigment that Borthwick and his colleagues had studied in the 1940s and 1950s (imaginatively labelled A to E) that sense light at the red end of the spectrum between 600 and 750 nm (Briggs & Olney, 2001). Two cryptochromes (1 and 2), which are blue-light receptors sensitive between 320 and 500 nm, are involved along with the phytochromes in the general entrainment of the circadian system to the daily light/dark cycle. A third set of photoreceptors, the phototropins, are also sensitive to light between 320 and 500 nm, but these pigments are not thought to be involved in the circadian system. There are almost certainly more photoreceptors to be uncovered.

The light signal entrains the rhythmic rise and fall in the abundance of proteins within the plant cells generated by a set of circadian clock genes. In *Arabidopsis*, the key genes and proteins involved in such a cycle are shown in Figure 3.10.

The gene–protein–gene loop of the *Arabidopsis* circadian clock is almost certainly more complex than that shown in Figure 3.10 and will involve more genes and proteins (Ueda, 2006). Recently, a team of Korean researchers has found a gene that regulates the period of the plant's clock. With a good sense of humour, they named the gene *FIONA1* after the heroine in the film *Shrek*. In the film, Princess Fiona is human by day but becomes an ogress at sunset. Fiona also sounds similar to the term 'flowering' in Korean (Kim *et al.*, 2008).

Bünning's hypothesis that the circadian rhythm that drove the leaf movements also drove the timing of the initiation of flowering and all the other photoperiodic responses has also been investigated at the molecular level. His idea (also known as the External Coincidence model, depicted in Figure 3.11) was that the timing mechanism had two alternating phases of

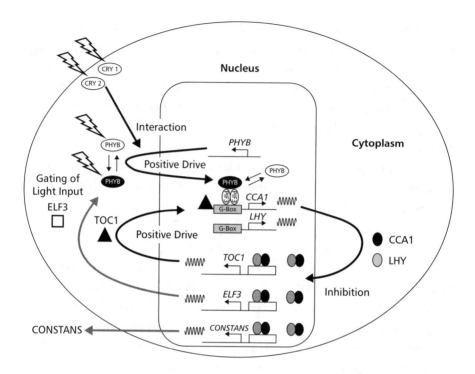

**Figure 3.10** The *Arabidopsis* circadian clock arises from a negative feedback loop. Adjustment (entrainment) of the *Arabidopsis* circadian clock to the dawn/dusk cycle is thought to occur through a combination of the actions of phytochromes (PHYB) and cryptochromes (CRY1 and CRY2). Phytochrome enters the cell nucleus after absorbing an activating photon of red light (converting it to the active far-red-absorbing form). In the nucleus, phytochrome interacts with the transcription factor PIF3. This complex (along with TOC1 – see below) then increases the transcription of genes containing the PIF3 DNA-binding site (G-box), which include the genes *LHY* and *CCA1*. Transcripts of the genes *LHY* and *CCA1* peak around the late subjective night/early dawn. The protein products of these genes, LHY and CCA1, are capable of binding directly to the promoter of the *TOC1* gene and inhibiting its transcription. Transcript levels for the gene *TOC1* show oscillations 12 hours out of phase with *LHY* and *CCA1*, peaking at the end of the late subjective day. TOC1 is thought to be a critical part of the positive element of the circadian oscillator (with PHYB and PIF3), acting by some means to increase the transcriptional rates of *CCA1* and *LHY*. These three proteins compose the central core elements of the *Arabidopsis* circadian clock. The circadian-regulated output protein, ELF3, is thought to feed back on to the clock input pathway, and, as a consequence, to gate or limit the clock response to light to particular times of day. Protein abbreviations: CCA1, Circadian Clock Associated 1; CRY1 and CRY2, cryptochromes 1 and 2; EL4, Early Flowering 4; LHY, Late Elongated Hypocotyl; PHYB, phytochrome (of which there are multiple types); TOC1, Timing of CAB Expression 1.

about 12 hours each, which he distinguished as a light-loving (photophilic) period during the day and a dark-loving (scotophilic) period in the dark. Light falling on a plant at a specific phase during the photophilic phase will enhance flowering initiation, and during the scotophilic period will inhibit it. In practical terms, Bünning's hypothesis stated that light has a dual function: light entrains a circadian clock, which in turn drives the rhythm of photosensitivity; light falling during the light-sensitive portion of the rhythm will trigger flowering. In other words, there is a specific, critical, photo-inducible phase, and when this interacts with light, seasonal events are set in train (Figure 3.11).

The near-24-hour rhythm generated by the circadian clock is synchronised by dawn and dusk, and this rhythm regulates the expression of perhaps a quarter of the genes in *Arabidopsis*. One of these genes, known as *CONSTANS*, has a key role in the transformation of the light-sensitive reactions of the photoreceptors into physiological activity by the plant, such as initiating flowering, particularly through its interaction with the *FT* (*FLOWERING LOCUS T*) gene, which induces a small group of undifferentiated cells to produce flowers at the tip of a plant stem (Figure 3.12).

George Coupland and his team have made direct measurements of the concentration of CONSTANS (CO), the protein expressed by the *CONSTANS* gene, in *Arabidopsis*. The integrity of the protein depends on its exposure to light. Without light, the protein is destroyed and its concentration remains low, even when its precursor mRNA is abundant. To reach high levels of CO protein, there must be an overlap between the clock-controlled *CO* mRNA and the light-driven response of the photoreceptors (Valverde *et al.*, 2004). So, in a long-day plant such as *Arabidopsis*, exposure to light in late afternoon or evening initiates high levels of CO. During short days, only low levels of CO will be produced. In this way, a seasonal daylength signal is introduced into the clock cycle that enables the plant to turn its 'reading' of the time of year into a physiological activity such as flowering.

The picture emerging in *Arabidopsis* shows how a long-day plant can discriminate short from long days and integrate the information into its developmental programme. However, our molecular understanding in other plants, particularly short-day species, is still at a very preliminary

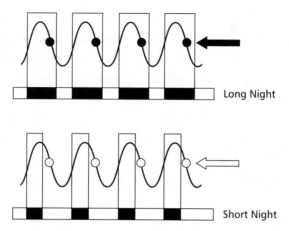

**Figure 3.11** Bünning's hypothesis or 'External Coincidence model' of photoperiodic regulation. An internal circadian clock tracks day and night length throughout the year and controls the level of a regulatory molecule whose activity is directly affected by light. During long nights the photo-inducible phase (black dots) is not exposed to light, but during short nights the photo-inducible phase (white dots) encounters light, triggering a photoperiodic response. Light both promotes long-day plant flowering and inhibits short-day plant flowering, and also sets the phase of the oscillator (Yanovsky & Kay, 2003).

stage and may not apply to all species (Ramos *et al.*, 2005). It is a sophisticated process and indicative of how millions of years of natural selection have resulted in plant species able to synchronise their life cycles exquisitely to a temporal and spatial environment. Plants deserve more respect! They have neither eyes, nor brain nor nervous system. Yet over the past hundred million or more years, the flowering angiosperms have developed a set of light-sensitive receptors and a signalling system that, linked to an innate biological rhythm, enables them to know both the time of day and the time of year.

However, there is a problem with using photoperiod alone, even among those plants that use the light signal. Even if the day or night length is predictable from year to year, the weather is not. As Martin Lechowicz of McGill University has explained (Lechowicz, 2001):

> Just knowing the time of the year is insufficient to evaluate the likely environmental conditions in the weeks or months ahead. Lengthening days can cue the onset of spring, but not whether the spring will advance quickly or slowly, or be

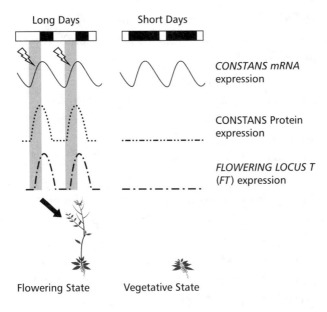

**Figure 3.12** Photoperiodic control of flowering in the long-day plant *Arabidopsis*. The diagram shows the clock-controlled expression of *CONSTANS* (*CO*) mRNA (messenger RNA) under long and short days. During short days, *CO* mRNA expression peaks during the night, but CO protein does not accumulate (in the absence of light); as a result the downstream gene *FT* is not expressed. Under long-day conditions (16 hours of light:8 hours of darkness), the peak of *CO* mRNA expression partly coincides with light, the protein accumulates in the nucleus and the expression of *FT* mRNA is activated. FT itself promotes early flowering (Figure 3.10) (Searle & Coupland, 2004). Many other genes and proteins are involved in regulating the cycle, and several important elements remain to be discovered, notably the mechanism that allows the photoreceptors to alter the function of CO.

especially warm or cool. Organisms thus use temperature, and sometimes water regimes, to better estimate the progression of seasons in a given year and locality. These climatic signals are useful cues for phenological events, but also not entirely reliable. Changes in atmospheric circulation may indeed bring an earlier than usual onset to spring-warming, but not necessarily preclude a late freeze.

Plants add in a variety of cues and adaptations to be on the safe side. Biennial plants in particular have to be careful to avoid being 'tricked' into flowering in the late autumn by a cold snap that is followed by warm conditions (Sung & Amasino, 2004). Vernalisation is very common among plants in temperate latitudes and it acts as an added 'fail-safe' system to the daylength measurement of photoperiodism to ensure flowering at the appropriate time of year. It comes from the Latin *vernus*, meaning 'of the spring', and a vernalised plant is one that will not flower without undergoing a prolonged period of cold. It is an important adaptation for plants that begin growing in one season and flower in the following spring because it enables them to distinguish spring from temperature fluctuations in the autumn. In the first growing season the plants become established, and in the next spring they flower rapidly to take advantage of favourable conditions and to avoid competition. In many species, vernalisation is not sufficient in itself to induce flowering but renders plants competent to flower when the photoperiodic signal occurs.

An example is henbane (*Hyoscyamus niger*), a biennial plant that was traditionally used in German pilsner beers as a flavouring, until the Bavarian Purity Law, passed in 1516, outlawed its use and allowed only hops. Henbane will not flower unless it has been exposed to a cold winter. Vernalised henbane plants kept in a short-day photoperiod grow vegetatively, but do not flower. The vernalised henbane plants readily flower, however, when shifted to long-day photoperiods. Effectively, they 'remember' the cold spell, and the lengthening days of spring allow the initiation of the flowering response to proceed, or perhaps remove an inhibition. In some species, the 'memory' lasts days; in others it can last years.

Because of vernalisation, winter cereals can be planted in the autumn to take maximum advantage of the favourable growing conditions in the spring (Sung & Amasino, 2004). Although vernalisation is a protective adaptation that guards against plants' initiating flowering too early in the year by somehow sensing and counting a period of cold temperatures, it does little to help the plant through the winter itself. As autumn fades, plants sense the falling temperatures of approaching winter and develop tolerance to the cold in a process known as cold acclimation. Unlike vernalisation, when up to 40 or more continuous cold days may be needed in some species to institute the process, most plant species that can acclimatise to cold do so after a day or so of low temperatures. Gene expression changes within minutes of exposure to cold.

Seasonal cold acclimation of biennial and perennial woody plants native to the temperate zones has been described by Leena Lindén in the following way (Lindén, 2002):

> [A] combination of the photoperiodic effect and an acute response to cold. The first stage is strongly affected by photoperiod. In many woody plants, shortening days of autumn induce growth cessation, which is a prerequisite for cold acclimation. Cells in the first stage of acclimation can survive temperatures well below $0°C$, but they are not fully hardened. The second stage of cold acclimation is induced by low temperature, especially subzero temperatures. During this stage plants undergo metabolic and/or structural changes, which lead to a considerable degree of cold hardiness.

The close relationship between the photoperiodic response and the geographical origin of plants suggests that photoperiodism is an adaptive response. Plants growing in temperate and high latitudes tend towards the formation of long-day species, and the plants of subtropics and tropics are directed towards the establishment of short-day species. Short-day and long-day species originated from more ancient day-neutral species, and the adaptation of the former species to length of day gave them advantages over neutral forms in the appropriate conditions. Plants compete with each other for resources, which includes sunlight, and any advantage that can be gained, say from flowering earlier or later, would be selected.

Long-day plants fail to flower under short autumn and winter days but instead produce small bushes and rosettes. This promotes their preservation under a snow cover and represents their adaptation to overwintering in temperate and northern latitudes. Short-day plants fail to flower under long summer days but continue vegetatively and survive hot and dry summers or periods of pouring rains in subtropical and tropical countries. But it must be emphasised that daylength is only an indirect cue to the anticipation of a season such as the arrival of winter. It is the timing of winter's advent, not daylength itself, that imposes selection pressure. So if the timing of the seasonal climate changes, as is happening with climate change, some plants that get their 'timing' wrong may benefit. Others will become extinct.

# 4

# SEASONAL REPRODUCTION IN MAMMALS AND BIRDS

Through the hush'd air the whitening shower descends,
At first thin-wavering; till at last the flakes
Fall broad, and wide, and fast, dimming the day
With a continual flow. The cherish'd fields
Put on their winter robe of purest white.

EXCERPT FROM 'WINTER', IN *THE SEASONS*, BY JAMES THOMSON
(WHITING & LEAVENWORTH, FOR THOMAS, ANDREWS &
PENNIMAN, 1801)

Animals, just as plants, time reproduction to fit closely with the changes in the seasons. In their case this means ensuring that the young are born when the maximum amount of food is available. No matter the species, the best-fed young are the most likely to survive to become breeding adults (Austin & Short, 1985), whether they be on land, in the sea or in the air.

For many herbivorous mammals, births are timed for spring, but for their predators the timing is synchronised to the development of their prey. Elk, deer and many other ungulates (hoofed animals) mate in the autumn in the northern hemisphere and the offspring are born in the following spring (March to May). The burst in the growth of grass at that time provides the mother with sufficient calories to produce enough milk for her offspring and for weaning the young. Predatory wolves time their breeding not to the abundance of grass but to the abundance of new and immature young animals. The cubs are suckled initially but begin to eat regurgitated meat after about two weeks of age and are fully weaned by 8–10 weeks (late June or early July) (Mech & Boitani, 2003). The mother, working as a member of the pack, will feed mainly on young or sick deer, moose, elk and bison. Because a typical adult female wolf requires a minimum of 1.1 kg of food per day for sustenance, but approximately 2.2 kg to rear cubs successfully, she is not too particular as to what meat she eats. Wolves will take advantage of the springtime burst in the population of small mammals to supplement their diet.

Timing reproduction is more complicated for predatory polar bears that live mainly on seals. Much of our understanding of their reproduction comes from studies on animals living in the Svalbard archipelago, midway between Norway and the North Pole. The testes of male polar bears are withdrawn into the abdomen for most of the year. They descend into the scrotum in late winter and remain there until May. Descent of the testes permits sperm production from February to May. In Svalbard, bears mate between March and June. Once mated, females begin depositing fat and need to gain a minimum of 200 kg for a successful pregnancy. They usually build a maternity den some 16 km from the coast in mid to late October, digging dens in snowdrifts on south-facing slopes. The tiny cubs, weighing in at about half a kilogram and 30 cm long, are born between November and January, and by late March they emerge from the den into less severe weather, although daytime temperatures can still fall below −25°C. By then the mother will have lost about 30 per cent or more of her body weight. She may be so hungry that she might even risk an attack on a young walrus. Male polar bears spend the winter outside on the sea ice and, given the chance, will kill and eat the newly emerged cubs.

Seals are the bears' main staple diet. Because seals cannot give birth in the water, they haul themselves out and give birth on the frozen ice shelf in early spring. Harp seals give birth to their white-coated pups on the unstable pack ice, where they are less likely to encounter a polar bear. Hooded seals also give birth on the more unstable parts of the ice shelf but are less vulnerable because they feed their pups on milk that contains 60 per cent fat, and weaning can be an incredibly short four days. Ringed seals, by contrast, breed farther north on the permanent sea ice. This makes them particularly vulnerable to attack by polar bears, so the pups are hidden in small ice caves. The polar bear uses its acute sense of smell to detect a seal in its ice cave as much as two kilometres away. Bears home in on the lair, where they succeed, on average, once in every 20 attempts to trap, excavate and eat the pup (Beeby & Brennan, 2003). After weaning her cubs in the spring of their third year, a female polar bear can then mate again during the same spring (Norris *et al.*, 2002).

Among birds, as with mammals, the overwhelming maxim for at least one of the parents is that the young must be fed! Rooks (*Corvus frugilegus*) rely heavily on earthworms to feed their young. They time the production of chicks to April and May, while the surface soil is still moist. As the weather becomes warmer, and the soil dries out, earthworms move deeper and become less accessible.

The grey partridge (*Perdix perdix*) pairs up from January to February, and most chicks are born in June. By this time the grass seeds and insects they feed to their young are in abundance. Finches, such as the goldfinches of North America, also feed their chicks on seeds and similarly time the production of their young to June. The Eleonora's falcon (*Falco eleonorae*) breeds very late from mid-July to September on uninhabited rocky islands in the Mediterranean and off coastal northwest Africa. Chicks begin to hatch in late August. The late breeding means that the nestlings are fed on small song birds migrating from Europe to Africa, which are successfully intercepted by the adult falcon parents (Lofts, 1970).

There is an optimal timing of birth that maximises the probability of survival of the offspring. Although the width of this timing window varies with species, getting it wrong, even by a small margin, reduces the chances of the offspring. But the availability of food itself is very seldom the proximate trigger for reproduction. Because there is a lag, of as much as a year or more in some mammals, between conception and birth, the question arises: How do animals correctly time conception so as to maximise not only future reproductive success in terms of birth but also the survival of the young into sexually mature reproducing adults?

With few exceptions, non-equatorial animals do not breed all year round. They save energy by effectively 'turning off' their reproductive organs for much of the year. In many species the gonads regress and in some they almost vanish. In the non-breeding state, the reproductive organs of many seasonally reproductive birds weigh no more than 0.02 per cent of body weight, but in full breeding condition the testes of the male can weigh between one and two per cent of total body weight (more than the weight of the brain). This would be a huge and unnecessary extra load for a bird dependent on flight.

Small mammals such as the Syrian (golden) hamster, which has a paired testicular weight of about 5 grams when reproductively active but only about 0.2 grams when in the non-breeding state, also regress their gonads. There is a relatively less marked seasonal change in the size of the reproductive organs in larger animals such as sheep and other ungulates. The scrotal circumference – a measure of paired testis size – of a ram increases from about 30 cm in the non-breeding state to about 36 cm when breeding. Even though the change in the gonad size in large mammals such as ungulates is fairly small, sperm production in the male and ovulation in the female is often shut down for many months of the year (Austin & Short, 1985).

Depending on the species, in most birds and mammals it takes between one and two months to reactivate a fully regressed set of reproductive organs. These physical changes are accompanied by behavioural changes, which frequently involve establishing and maintaining a breeding territory to attract a mate. After courtship and copulation, yet more time is needed for the young to develop before they can emerge into the world. This period of development, between fertilisation and birth or hatching, is largely fixed for each species. The whole business of procreation can take a couple of months for a bird or small mammal, to over a year in a large mammal such as a horse and nearly two years in elephants. If a peak in food availability were the trigger for reproduction, it would be long gone before the young arrived. So, to return to the key question, what triggers seasonal reproduction?

This is the same question in principle as that asked by Garner and Allard about plants in the second decade of the last century. Just as Allard and Garner had their predecessors among botanists, so zoologists had also been posing the question in the nineteenth century. Alexander von Homeyer, a

German ornithologist, suggested in the 1880s that daylength might control annual cycles, and similar proposals were made by the British physiologist Edward Albert Schäfer in the early 1900s (Schäfer, 1907).

In the 1920s, William Rowan provided the first experimental evidence for a photoperiodic response in birds. Rowan was born in Switzerland but spent his early life in Britain and then moved to Edmonton, Alberta, in Canada. He observed the migration of the greater yellowlegs (*Tringa melanoleuca*) for 14 years. The bird breeds in Canada, migrates south to Patagonia in the autumn and returns in spring, a round trip of some 26,000 km. The eggs all hatch between 26 and 29 May, and Rowan wondered what could regulate such a precisely timed series of events (Rowan, 1925). He considered environmental cues such as temperature and food availability, but concluded that only the change in daylength could provide such precision. He tested this hypothesis by capturing dark-eyed juncos (*Junco hyemalis*). These Juncos breed in Canada and overwinter further south, and Rowan caught them during the autumn migration. The birds were kept in sheds and exposed to artificial spring-like daylengths. Despite the sub-zero temperatures of the Canadian winter, the birds could be triggered to breed by the artificially long days (Rowan, 1929).

Although secondary factors such as a sudden cold spell or a warm spell might speed up or slow down the rate of reproductive development, temperature or food availability by itself does not trigger reproductive development in many temperate bird and mammalian species. As Rowan found, the daylength-driven photoperiodic response is the key timing trigger. Rowan considered that there might also be some other internal timer or 'physiological rhythm' in migrant species that experience complex and potentially confusing photoperiods at low latitudes or in the southern hemisphere. In this regard Rowan anticipated the discovery of circannual rhythms in birds some 30 years later by Gwinner (Gwinner, 1996b).

Farmers and animal breeders have known all about the effects of daylength for centuries. More than 200 years ago, Spanish farmers were providing artificial illumination at night to increase the egg production of chickens (Lofts, 1970), a common practice in the poultry industry of today. Domestic chickens are still very mildly photoperiodic despite the very best efforts to breed chickens that lack this trait. Thoroughbred racehorses in the northern hemisphere all technically become a year older on 1 January (1 August in the southern hemisphere). This determines whether they can race as one-year-olds, two-year-olds, or whatever. Two horses, one born in June and the other in January of the same year, are considered to be the same age even though the horse born in January will be six months older and have a significant advantage in terms of physical development. A horse born in December 2008 will be a two-year-old on 1 January 2009! Consequently, breeders want the mares, which have an 11-month gestation, to conceive in February or March and to foal in January or February. But mares naturally conceive in late spring or early summer and foal in the following late spring. Mares are fooled into conceiving early by keeping the light on in their stalls at night so they think the days are lengthening.

Following on from Rowan's findings, in 1933 John Randal Baker, a zoologist at Oxford University, went on an expedition to the New Hebrides in the Pacific to study the breeding seasons, if any, of the birds and animals of the rainforest in a climate of great uniformity, and relate them to any corresponding changes in their environments (Willmer & Brunet, 1985). He and his colleagues found that in this tropical area, breeding may occur in almost any month, but each species had its own period. For instance, in spite of the extremely uniform conditions, the breeding season of permanently cave-living bats in the New Hebrides was confined to a few weeks in early September (early spring at that latitude).

The timing cues used by species in the tropical zones are often unclear, but they are known in a few species. The red-billed weaver bird (*Quelea quelea*) of tropical Africa uses rainfall and the growth of new grass to trigger growth of the reproductive system. Fresh flexible grass stems are of paramount importance to this bird, which builds its remarkably elaborate woven nest from the new growth of grass. Animals living at the equator, where there is little change in daylength, were thought to show no photoperiodic response. However, if weaver birds are exposed to artificially long photoperiods, they will be stimulated to breed, suggesting that they, and perhaps many equatorial species, have some capacity to respond to daylength (Griffin, 1964; Hau *et al.*, 1998).

The changing daylength is used to time reproduction in a vast number of species, but the precise daylength selected varies greatly. Birds and small

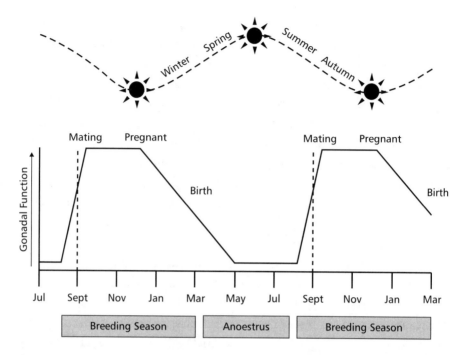

**Figure 4.1** The photoperiodic control of reproduction in sheep. The decreasing daylengths of late summer (August and September) trigger reproductive development. Ewes mate in autumn, and birth of the lambs occurs in early spring. For much of the spring and summer, ewes show no oestrous cycles (they are in anoestrus).

mammals with a relatively short period of development, either in an egg or in the uterus, often use the increasing daylengths after 21 December. Larger mammals such as sheep and deer have a pregnancy that varies between five and nine months. Their breeding is triggered by the decreasing daylengths of autumn. The rut and mating occur between September and December, and the young are born in the following spring or early summer (Figure 4.1). Some mammals have a very long gestation. They are also stimulated to breed by the increasing daylengths after 21 December. Mating occurs in the spring and summer, with birth nearly a year later.

Although the development time of the young in the uterus is fixed in most mammals, some use a remarkable trick to increase the apparent gestation period. Embryonic diapause, also known as discontinuous development or delayed implantation, effectively uncouples the timing of mating and fertilisation from birth by maintaining the embryo in a state of suspended development. Embryos produced at mating develop only as far as a hollow ball of cells (the blastocyst) before developmental arrest. This strategy is usually employed to ensure that birth and postnatal development occur under favourable environmental conditions. Although polar bears mate between March and June, the blastocyst stops growing and lies free-floating in the uterus for about four months. After implantation, there is a gestation of some four months, with birth between November and January. By the time the mother and cubs emerge from the den in late March, the weather conditions are less severe.

The common harbour seal is largely solitary; the only real social contact is between the mother and her nursing pup. The sexes meet once a year to mate, which occurs shortly after delivery of the pup from the previous mating. Because the harbour seal has a gestation of seven to eight months, embryonic diapause enables the female seals to make up the three-month difference and enable birth, the return of oestrus and then mating at the same time each year. The development of the egg after its arrest at the blastocyst stage depends on secretions from the wall of the uterus. These secretions are regulated by the ovary, which is in turn regulated by the hypothalamus in the brain. Shortening daylengths stimulate the production of the secretions, which in turn break the arrested development and trigger implantation.

In marsupial mammals, including the kangaroo and wallaby, there is an added refinement. When a female gives birth she also becomes receptive and mates. Embryos produced at this mating develop only as far as a blastocyst. Further development is blocked by the hormone prolactin, which is produced in response to the sucking stimulus from the newborn joey in the pouch. When sucking decreases as the joey begins to eat other food and to leave the pouch, or if the young is lost from the pouch, the blastocyst resumes development.

The photoperiodic responses that trigger the hormonal and behavioural (neuroendocrine) changes involved in reproduction are extremely sensitive. A difference of only 8–10 minutes in daylength can initiate reproduction in some species. At the Arctic Circle, this 10-minute change in daylight hours can take place over a two-day period in late summer. So, if it is daylength that enables animals to know the time of year and synchronise conception and reproduction well in advance, how do they measure it?

As in plants, two models have been proposed for the light induced photoperiodic response: (1) an 'hourglass-like' timer and (2) the Bünning hypothesis or External Coincidence model. But because biology is complicated and species vary so much, an additional model, called the Internal Coincidence model, was suggested by Colin Pittendrigh (Figure 4.2). He originally proposed this third model to account for insect behaviour, but more recent work on the reproductive patterns of sheep and other large mammals has endorsed its explanatory potential in these species (this is discussed further below).

To recap the explanations in the previous chapter, the hourglass timer is based on the idea that an unknown substance, which we can call X, is converted to substance Y in the light, and then in the dark Y is converted back to X. At the appropriate daylength, substance Y will reach a critical concentration or threshold and the seasonal event will be triggered (Figure 3.7).

Both the External and Internal Coincidence models link photoperiod with circadian rhythms. The External Coincidence model is based on the concept that it is not the total duration of light exposure that matters, but the point or phase at which the light occurs after dawn that is critical. Light at dawn and dusk entrains a circadian oscillator that drives a rhythm in light sensitivity. When light falls during a photo-inducible phase, the photoperiodic response is triggered. The difference between this concept and the hourglass model is that the animal or plant does not need to actually experience a full 14, 15 or whatever hours of light to trigger reproductive events; it simply needs to experience light at dawn and then light approximately 14, 15 or whatever hours later (a 'skeleton photoperiod'). The time between can be spent in darkness. It is the timing of the light pulse rather than its duration that is critical; a very short light pulse can operate a photoperiodic 'switch' if it falls at the correct phase of the rhythm's cycle. Brian Follett and Peter Sharp demonstrated this in the late 1960s while working on the mechanics of photoperiodism in birds at the University of Bangor in North Wales (Figure 4.2). They exposed Japanese quail to different skeleton photoperiods. Starting at dawn, all the birds received six hours of light, and then different groups of birds were left in various periods of darkness before being

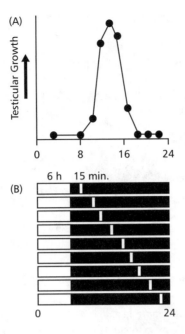

**Figure 4.2** (A) A 'skeleton photoperiod' showing the effects of 15 minutes of light exposure at different times after dawn on the rate of testicular growth in Japanese quail maintained under different 6-hour and 15-minute photoperiodic days. (B) Diagram indicating the timing of the 15-minute pulse of light. The precise position of the photo-inducible phase varies from species to species. In the quail, the peak of sensitivity is some 14 hours after dawn, whereas in other birds such as house finches it is 12 hours. For most birds, in winter the short days and long nights mean that the photo-inducible phase falls during darkness, but as the days lengthen the photo-inducible phase will be exposed to light, and seasonal events will be triggered.

exposed to 15 minutes of light. The cycle was then repeated another 13 times for each group to give a total of two weeks. In each case the birds were exposed to a total of 6 hours and 15 minutes of light in each cycle. However, in this species, it was only if the 15-minute pulse of light was between 12 and 16 hours after dawn that the birds were stimulated to breed, as measured by the amount of testicular growth and reproductive hormones they produced (Follett & Sharp, 1969). Skeleton photoperiod experiments in mammals have similarly suggested that a circadian timer sits at the core of photoperiodic time measurement, and such findings are entirely consistent with an External Coincidence model of photoperiodism (Yasuo *et al.*, 2003)

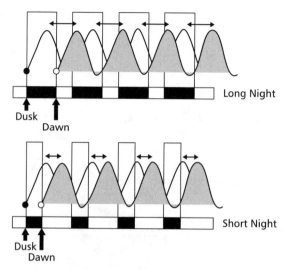

**Figure 4.3** Internal Coincidence model for photoperiodic regulation. In this model there are two circadian oscillators: one set to dusk (evening oscillator) and the other to dawn (morning oscillator). Photoperiodic induction is regulated by the phase relationship between the two sets of oscillators. Under long nights the oscillators are held apart, whereas under short nights the oscillators move together and overlap with coincident phase points.

Pittendrigh's alternative model suggested that the photoperiodic clock might comprise two oscillators, one entrained by dawn and the other set by dusk. As the photoperiod changes with the season, the phase relationship between the two oscillators changes, and this encodes daylength (Figure 4.3). In his model, light has a single entrainment role rather than both the entrainment and photo-induction of the External Coincidence model.

Distinguishing between an hourglass timer and the rhythmic timer of both the External and Internal Coincidence models has been relatively straightforward in most vertebrates, although the invertebrates are a problem, as we shall see in the next chapter. The experimental protocols, including the use of skeleton photoperiods, are described in Appendix I, but in essence animals are exposed to a light/dark cycle in which the light period is 6 to 12 hours and the dark period is between 18 and 72 hours. Photoperiodic induction occurs in plants, birds and mammals when the total period length is close to 24 hours or multiples of it. Periods such as 36 or 60 hours are non-inductive. This indicates the involvement of an endogenous 24-hour circadian pacemaker rather than an hourglass timer.

There is a potential problem for birds if the critical timing 'window' comes early in the spring, say in early April. They will be triggered to lay their eggs and the young will arrive several weeks later, while the daylength is still increasing (until 21 June). If the adults then breed again, a new batch of young would eventually be born when food supplies will have dwindled or disappeared. In these circumstances, reproduction has to be turned off so that another clutch of eggs is not laid – the bird must be made insensitive (refractory) to the formerly stimulatory photoperiod.

Just like the timing of the onset of reproduction, the timing of its termination differs greatly between birds, and these differences depend on the birds' responses to photoperiodic information. For instance, an 11-hour daylength in early spring triggers reproduction in the British starlings. The starling lays its eggs but normally does not copulate again in the same year. Only six weeks from the triggering of reproduction, and while the daylength is still increasing, the reproductive system of British starlings collapses. Not far away, in the hills of southern Germany, starlings studied by Helga Gwinner at the Max-Planck-Institut für Ornithologie, are regularly double-brooded. Such differences between the populations of the same species in the timing of the collapse of the reproductive system have been described for many birds, and these differences highlight the importance of local conditions in the evolution of seasonal breeding patterns.

Some single-brood birds such as the British starlings are insensitive (refractory) to long days after breeding and undergo reproductive collapse. If kept artificially in long days they will never breed again. But in nature the system is reset by exposure to the shortening daylengths of the autumn so that the bird will once more respond to the increasing daylengths in the following spring. The collapse of the reproductive system is regulated by a complex set of hormonal interactions involving the pituitary hormone prolactin and hormones secreted by the thyroid glands (Dawson et al., 1986). Other single-brooded species, such as the garden warbler (Sylvia borin), regain reproductive competence even under long (16L:8D) days. In these long-distance migrants, photorefractoriness is dissipated gradually under the control of a circannual timer (Gwinner, 1986).

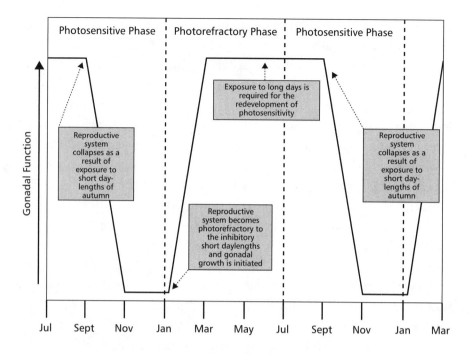

**Figure 4.4** The effects of daylength on reproduction (gonad function) in a small mammal such as a golden hamster. Testis regression occurs in late summer or autumn as a result of exposure to shortening daylengths. 'Spontaneous' development (recrudescence) of the gonads occurs in early spring because the hamster has become insensitive (refractory) to the previously inhibitory effects of short days. If kept on short daylengths the hamster will remain reproductively active for the rest of its life. Exposure to long days is required for short-day-induced gonadal regression.

Hamsters and other small mammals regress their reproductive system in the autumn in response to the shortening days. The reproductive system remains regressed, but after the winter solstice the reproductive system starts to develop again spontaneously. Exposure to long days is then needed to maintain an animal's reproductive state and enable it to become sensitive once more to the inhibitory effect of the short days of autumn. If a hamster is kept in an artificial environment of short daylengths, it will remain insensitive to this photoperiod and remain reproductively active for the rest of its life (Figure 4.4).

Sheep have a marked seasonality of breeding activity (Figure 4.1). The photoperiod is the determining factor of this phenomenon, and temperature, nutritional status, social interactions, lambing date and lactation period modulate it (Lincoln, 1998). In tropical zones, where daylength remains relatively constant, ewes tend to remain sexually active throughout the year. In temperate areas, ewes enter a non-reproductive (anoestrous) state during the spring and summer, and start reproductive cycling in the autumn as the daylength decreases. The autumn breeding is not actively driven by decreasing or short daylengths, but rather because the sheep become insensitive to the inhibitory effect of the long days.

Although the precise photoperiodic mechanisms in mammals and birds vary in detail, suggesting a long and diverse evolutionary history, the general principle of a circadian-based timer is much the same. The driving force of the mammalian circadian molecular clock, as in plants, is based on a transcriptional/translational feedback loop (Figure 4.5), but different genes and proteins generate this molecular oscillation (Foster *et al.*, 2004). The system is complex, but at its heart the feedback loop sets up a near-24-hour oscillation in protein abundance and degradation, and this is the basis for a circadian rhythm. The detail shown in Figure 4.5 is complicated, as it was for plants in the previous chapter, but readers who have a basic knowledge of gene activity and of transcription and translation should be able to make sense of what is happening. This model is, however, more detailed than is needed for general understanding.

The master circadian pacemaker in mammals is in a small paired structure called the suprachiasmatic nuclei (SCN) in the anterior hypothalamus of the brain. The SCN has some 20,000 neurones, which individually are capable of generating near-24-hour oscillations in electrical activity. Most mammalian tissues also express the genes shown in Figure 4.5 and have the capability of generating a circadian rhythm in clock-gene proteins. For example, liver cells can be removed and isolated in culture and will show 24-hour oscillations in gene expression for several cycles before damping out. The role of the SCN is to coordinate the activity of these 'peripheral' oscillators in the body. The SCN acts a bit like the conductor of an orchestra, coordinating the multiple rhythmic parts of the body to produce a properly timed and coordinated response (Foster & Kreitzman, 2004).

The rhythms in gene expression and protein production and degradation are constrained by zeitgebers (time-givers) into a biologically useful

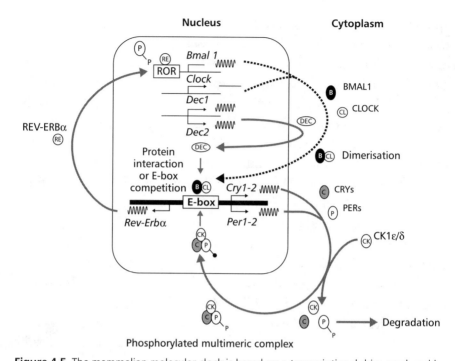

**Figure 4.5** The mammalian molecular clock is based on a transcriptional drive produced by CLOCK:BMAL1 heterodimers. The gene *Clock* is expressed continually, whereas *Bmal 1* is expressed rhythmically. The CLOCK:BMAL1 heterodimers bind to the E-box of the period and cryptochrome genes, producing rhythmic expression of *Per* and *Cry*. The resulting PER proteins are phosphorylated by CK1ε/δ. After phosphorylation, PER is either degraded or it interacts with CRY proteins to form a phosphorylated multimeric complex. This complex enters the nucleus and produces a negative feedback by inhibiting CLOCK:BMAL1-mediated transcription. An additional loop is produced via *Rev-Erbα*, which also possesses an E-box enhancer activated by CLOCK:BMAL1. REV-ERBα acts via a ROR element in the *Bmal 1* gene to inhibit *Bmal 1* transcription, thus feeding back to remove the positive drive produced by CLOCK and BMAL1. As the PER/CRY/CK1ε/δ complex enters the nucleus and inhibits the CLOCK:BMAL1 drive on the E-box, *Rev-Erbα* expression is also reduced. This leads to a disinhibition (activation) of *Bmal 1*, thus restarting the molecular cycle. *Dec1* and *Dec2* may modulate the CLOCK:BMAL1 drive by competing for E-box binding or sequestering BMAL1. Grey lines indicate inhibitory pathways on transcription; black dotted lines indicate pathways that lead to a transcriptional drive (Foster *et al.*, 2004).

precise 24-hour cycle that mirrors the daily solar cycle. Without this signal, the circadian rhythm will free-run at periods that can vary from 24 hours by perhaps 2–3 hours, depending on the species. The vital information that synchronises the circadian clock to the solar cycle comes from the dawn and dusk light signal which reaches the mammalian SCN via the eye, but not via the rods and cones. Instead there is an entirely separate and ancient photoreceptor system composed of a small number of photosensitive retinal ganglion cells (pRGCs) that use a photopigment, melanopsin (also called Opn4), which is sensitive to blue light (Foster & Hankins, 2007).

The sensitivity of pRGCs to the light signals of dawn and dusk enables mammals to synchronise with the sun's daily cycle. But they also have to know the time of year, and light is the connection. If the SCN of a photoperiodic mammal is damaged, the animal is incapable of using daylength changes to generate a photoperiodic response, so there must be an intimate link between the 24-hour circadian system and photoperiodic timing. The input of the dawn/dusk cycle detected by the pRGCs is encoded by the SCN into a photoperiodic signal that triggers the timing of the seasonal changes in physiology and behaviour so as to maximise reproductive success. A key breakthrough in understanding this process came from studies on the pineal gland and the hormone melatonin.

During the 1970s and 1980s researchers had found that removing the pineal gland in the golden hamster resulted in a reproductively active individual that was no longer dependent for its timing on the photoperiod (Reiter, 1975). Even disconnecting the pineal from its neural connection, which is supplied by the sympathetic nervous system, prevents mammals from distinguishing between long and short days. Bruce Goldman (now at the Department of Biobehavioral Sciences, University of Connecticut) and his colleagues then established that melatonin released by the pineal is the critical downstream link between the 'clock' and the reproductive system in mammals.

They removed the pineals from hamsters and divided the animals into two groups. One group received an infusion of melatonin into the blood that corresponded to winter (long-duration infusion), whereas the other group were infused with melatonin that corresponded to spring and summer (short-duration infusion). Those hamsters that received the spring and summer pattern of melatonin were reproductively active, but the winter melatonin pattern did not stimulate reproduction (Figure 4.6) (Goldman et al., 1984).

Complementary results were found in ewes, which are short-day

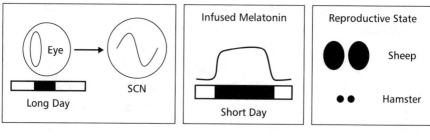

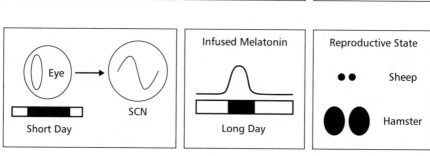

**Figure 4.6** Pineal melatonin encodes night length. This figure summarises the experiments that demonstrated the critical role of melatonin as the photoperiodic hormone. Pinealectomised sheep (short-day breeders) or hamsters (long-day breeders) were exposed to either a winter or a spring photoperiod but were infused with the opposite melatonin profile. The effects of this mismatch on the reproductive system was then determined. In every case the animals showed a reproductive response that corresponded to the pattern of infused melatonin and not the photoperiod.

(long-night) breeders. Pinealectomised ewes were exposed to a short-night (spring) light cycle and infused with a long-duration (winter) melatonin pattern or exposed to a long-night (winter) light cycle but infused with a short-duration (spring) melatonin pattern. The reproductive system of the pinealectomised ewes ignored the photoperiod, responding only to the pattern of infused melatonin. The winter melatonin pattern stimulated reproduction, but the spring profile left the animals unstimulated (Figure 4.6) (Wayne *et al.*, 1988).

During daylight hours there is almost no melatonin in the blood, and its release from the pineal mirrors the length of the night: long in the winter and short in the spring and summer. Destruction of the SCN blocks photoperiodic responses, so there has to be a way by which the SCN drives this melatonin signal – but how is it done?

The SCN regulates melatonin release from the pineal by a complicated series of relays through the brain and sympathetic nervous system. Essentially, the neurotransmitter noradrenaline (also known as norepinephrine) is released from the sympathetic nerve fibres in large amounts during the night in response to changing patterns in electrical activity in the SCN. Intrinsic electrical activity in the SCN is high during the day but low during the night. This oscillation in electrical activity continues under constant conditions of continuous light or dark, but in a light/dark cycle it is both locked onto (entrained by) and modified by light. The SCN signals dusk with a decrease in electrical activity, which remains low until dawn; then it increases once more. If a mammal such as a hamster is exposed to either long or short days for several weeks, and then its SCN is removed and its electrical activity is monitored in isolation from the rest of the brain, the SCN 'remembers' the daylength (VanderLeest *et al.*, 2007).

The pattern of noradrenaline release is an inverted image of the electrical activity in the SCN, so that raised electrical activity in the SCN decreases noradrenaline release from sympathetic nerve terminals in the pineal, and vice versa. During the night, there is a decline in the electrical activity of the SCN and a rise in noradrenaline levels. Receptors in the pineal cells bind noradrenaline and this ultimately increases calcium levels in the melatonin-producing cells. The calcium signal causes the activation of the enzyme arylalkylamine *N*-acetyl transferase (AA-NAT). This is the rate-limiting enzyme in melatonin production and so the rhythmic regulation of AA-NAT by the oscillatory SCN electrical activity determines the profile of melatonin production (Klein *et al.*, 1983).

Although it became clear that the profile of melatonin released from the pineal encodes the daylength signal and that the melatonin profile alone can regulate the cascade of hormonal activity of the reproductive system, the question became one of how it was done. What was the link between photoperiod, melatonin and the reproductive system?

The link is in the pituitary gland. The pituitary is an endocrine gland that regulates much of mammalian physiology and behaviour. It has two lobes, anterior and posterior. The anterior pituitary is further divided into anatomical regions known as the pars tuberalis (PT), pars intermedia (PI) and pars distalis (PD). There is a very high density of melatonin receptors in the individual cells of the PT. Melatonin binds to these receptors, which

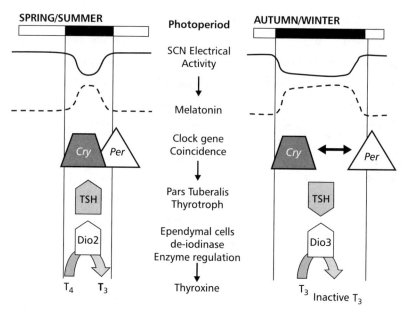

**Figure 4.7** Schematic diagram of the photoperiodic regulation of hypothalamic levels of thyroxine (T$_4$) in a sheep. The photoperiod, detected by photosensitive retinal ganglion cells (pRGCs), alters the electrical activity within the suprachiasmatic nuclei (SCN). Light increases neuronal activity, whereas dark decreases activity. This, in turn, regulates the sympathetic input to the pineal gland and melatonin synthesis and release. Melatonin mirrors the night length. The pars tuberalis (PT) of the anterior pituitary gland possesses a very high density of melatonin receptors, and binding of melatonin to these receptors alters the expression of the *Per* and *Cry* genes. *Cry* gene expression tracks melatonin rise, whereas *Per* gene expression tracks melatonin decline. As a result the interval between the peak of *Per* and *Cry* gene expression varies as the melatonin signal expands and contracts with varying photoperiod. This decoding of the melatonin signal resembles the Internal Coincidence model for photoperiodic timing. *Per/Cry* gene coincidence regulates the release of thyroid-stimulating hormone (TSH) from thyrotroph cells within the PT. TSH then acts on the ependymal cells of the hypothalamus to regulate de-iodinase enzymes. Under the increasing daylengths of spring, release of TSH from the PT is high and stimulates Dio2 to catalyse the conversion of T$_4$ to T$_3$. Under the decreasing daylengths of autumn, release of TSH from the PT is low, which allows Dio3 to catalyse the conversion of T$_3$ to an inactive form of T$_3$.

alters the gene expression of several of the clock genes, including the *Per* and *Cry* genes (Figure 4.5) within the cells. This altered pattern of gene expression is best understood in sheep (Figure 4.7). *Cry* gene expression tracks melatonin rise (dusk), whereas *Per* gene expression tracks melatonin decline (dawn). As a result, the interval between the peaks of *Per* and *Cry* gene expression varies as the melatonin signal expands and contracts with varying photoperiod (Lincoln *et al.*, 2003; Morgan & Hazlerigg, 2008).

The altered pattern of *Per/Cry* gene expression in the PT regulates many neuroendocrine events, including the metabolism of thyroid hormones within the brain. Until very recently, thyroid hormones were thought to be important in seasonal physiology, because they regulate gross metabolic activity, but not central to the photoperiodic process itself. But now it seems that these hormones are actually an integral component of the photoperiodic mechanism of both birds and mammals.

The link between thyroid hormones and photoperiodism was first made using animals whose thyroids had been removed. Thyroidectomy blocks the normal photoperiodic response of both sheep and hamsters, whereas thyroxine hormone replacement allows a normal response (Billings *et al.*, 2002; Barrett *et al.*, 2007).

Thyroxine production is regulated by thyroid-stimulating hormone (TSH) released by the cells in the PT (thyrotrophs). This process is itself regulated by thyroid-stimulating hormone-releasing hormone (TSH-RH; also abbreviated as TRH) from neurosecretory cells in the hypothalamus. The circulating form of thyroxine released from the thyroid gland is $T_4$ (thyroxine). This hormone has little biological activity, but if $T_4$ is stripped of an iodine (de-iodinated) at target tissues by the activity of the de-iodinase enzyme Dio2 it becomes the biologically active $T_3$ form. Another de-iodinase enzyme, Dio3, catalyses the conversion of $T_3$ to an inactive form of $T_3$ (Lechan & Fekete, 2005).

The vital link in the complex causal chain is the cells in the PT (thyrotrophs) that bind melatonin, which in turn alters the internal coincidence of *Per/Cry* gene rhythms in these cells. This regulates TSH release. TSH then acts on the ependymal cells in the hypothalamus to regulate the levels of $T_3$ through the Dio2 and Dio3 enzymes. There is a photoperiod-dependent change in Dio2 and Dio3 activity in the ependymal cells. During the increasing daylengths of spring, release of TSH from the PT is high and stimulates Dio2 to catalyse the conversion of $T_4$ to $T_3$. In the decreasing daylengths of autumn, release of TSH from the PT is low, which stimulates Dio3 to catalyse the conversion of $T_3$ to the inactive form. As this is true for both long-day-breeding hamsters and short-day-breeding sheep it seems

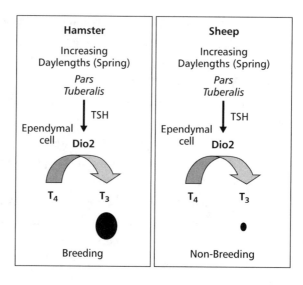

**Figure 4.8** The relationship between photoperiod, T₃ synthesis and reproductive status in the hamster and the sheep. In both the hamster and sheep, increasing daylengths promote TSH secretion from the pars tuberalis of the anterior hypothalamus. TSH travels to the cells lining the ventricle (ependymal cells) of the ventral hypothalamus. These ependymal cells show a photoperiod-dependent change in Dio2 activity such that under long days $T_4$ is converted to $T_3$. As this is true for both long-day-breeding hamsters and short-day-breeding sheep it seems that elevated levels of $T_3$ stimulate reproduction in the hamster but suppress reproductive activity in the sheep.

that elevated levels of $T_3$ stimulate reproduction in the hamster but suppress reproductive activity in the sheep (Figure 4.8). The effect of these changes on the seasonal breeding patterns in sheep, which are short-day breeders, and hamsters, which are long-day breeders, can be followed sequentially in Figures 4.7 to 4.10.

Although it is not yet certain, $T_3$ probably regulates a group of neuro-secretory cells in the hypothalamus known as gonadotrophin-releasing hormone (GnRH) neurones that project to the base of the brain (median eminence). These GnRH cells release the neurohormone GnRH in pulses into a small group of blood vessels (portal blood supply) that lead to the anterior part (pars distalis, or PD) of the pituitary. GnRH stimulates cells within the PD to release luteinising hormone (LH) and follicle-stimulating hormone (FSH). These hormones travel in the blood to the reproductive organs, where they stimulate reproductive activity and the release of testo-sterone and oestrogen. GnRH is released in seasonally modulated pulses, and the pituitary effectively 'reads' this time-based signal and releases LH and FSH in response to this pattern.

Precisely how the $T_3$ signal changes the pulse pattern of GnRH neu-rones is still unknown. The effect of the seasonally linked change in the pulse frequency is different in sheep and in hamsters. In sheep (short-day breeders) there is a high pulse (stimulatory) frequency of GnRH during the short days of winter, which ultimately triggers the reproductive system. However, during the long days of spring and summer, the pulse frequency of GnRH is long, and so the pituitary gland is not stimulated to release the reproductive hormones. In hamsters (long-day breeders), there is a high pulse (stimulatory) frequency of GnRH in summer and a low pulse (non-stimulatory) frequency in winter.

The photoperiod/melatonin signal acting on the pars tuberalis (PT) also regulates the release of prolactin from the pars distalis (PD). Prolactin regulates milk production and is also involved in regulating other aspects of seasonal physiology, including food intake, changes in metabolic rate, and winter coat growth (Lincoln et al., 2003). The precise mechanisms again remain unclear but the working hypothesis is that the PT responds to the melatonin signal, which drives the internal coincidence of Per/Cry gene rhythms and then regulates the production of a still to be isolated local prolactin-releasing factor that has been termed 'tuberalin', which trav-els to the PD and stimulates the release of prolactin (Figure 4.9). Figure 4.10 summarises the current description of the seasonal breeding pattern in sheep; although the details may well differ, it is probably a reasonable model for other mammals.

Although there is still uncertainty as to the precise way in which the photoperiodic melatonin signals are encoded and decoded at a molecular level, the broad picture of how the photoperiodic response synchronises mammalian reproduction to seasonal timing is making sense. However, it should come as no surprise to learn that the mechanisms that regulate sea-sonal physiology in birds are markedly different. The pineal of most birds contains a self-sustained circadian oscillator that regulates the release of melatonin from the gland. When the pineal is removed and maintained in

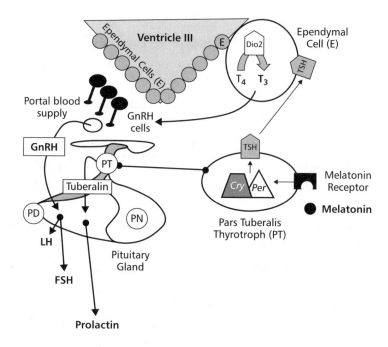

**Figure 4.9** The photoperiodic mechanisms regulating the release of pituitary hormones in the mammal. Changes in daylength alter the pattern of melatonin release from the pineal. Melatonin receptors on pars tuberalis (PT) thyrotroph cells alter the coincidence of *Cry* and *Per* rhythms, which in turn regulate the release of thyroid-stimulating hormone (TSH). TSH travels to the ependymal cells of the third ventricle of the brain and alters de-iodinase enzyme activity. Under increasing daylengths, Dio2 enzyme activity increases and leads to elevated levels of $T_3$. $T_3$ is then thought to alter the activity of gonadotrophin-releasing hormone (GnRH) neurosecretory cells within the hypothalamus. GnRH travels to the anterior pituitary gland and regulates the pattern of release of luteinising hormone (LH) and follicle-stimulating hormone (FSH) from the pars distalis (PD). The PT responds to the melatonin signal in another manner by producing a yet to be identified prolactin-releasing factor called 'tuberalin', which then travels to the PD and stimulates the release of prolactin. It is also likely that photoperiod modulates hormone release from the posterior pituitary gland (pars nervosa, PN), but so far there is no knowledge of how this might occur.

culture medium in a dish, the pattern of melatonin release is determined by the length of the dark portion of a light/dark cycle. Furthermore, when the gland is placed in darkness, the pattern of melatonin release 'remembers' the pattern of release under the previous light/dark cycle. So the avian pineal can autonomously encode the length of the night in terms of melatonin

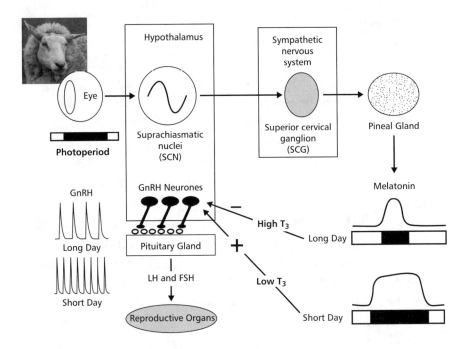

**Figure 4.10** Diagram summarising the photoperiodic regulation of reproduction in the sheep, showing the projections to the pineal gland from the eye, the suprachiasmatic nuclei (SCN) and the superior cervical ganglion (SCG). The photoperiod (night length) determines the duration of melatonin release from the pineal. When the days are short, the duration of nocturnal melatonin is long and this will stimulate short-day breeders such as sheep. The reverse is true when the days are long. The pineal melatonin signal ultimately determines hypothalamic levels of thyroxine ($T_3$) (Figure 4.9). Long days promote $T_3$ levels in the hypothalamus, whereas short days decrease levels of hypothalamic $T_3$. Levels of $T_3$ alter the activity of gonadotrophin-releasing hormone (GnRH) neurosecretory cells within the hypothalamus, but we currently do not fully understand how. The GnRH cells project to the base of the brain (median eminence), and GnRH is released into a small group of blood vessels (portal blood supply) that lead to the anterior part of the pituitary. GnRH is released in pulses. The pulse interval is decoded by the pituitary, and this alters the pattern of release of luteinising hormone (LH) and follicle-stimulating hormone (FSH). In sheep (short-day breeders) there is a high pulse (stimulatory) frequency of GnRH during short days, and this ultimately triggers the reproductive system. But when the days are long, the pulse frequency of GnRH is long, and so the pituitary gland is not stimulated to release the reproductive hormones.

release. However, unlike the situation in mammals, removal of the pineal has little or no effect on the ability of birds to display seasonal physiology! Although the melatonin signal of birds reflects night length, it does not seem to be used as such an indicator in a seasonal biology.

The photoperiodic response is lost in birds if an SCN-like structure within the medial-basal hypothalamus (MBH) in the brain is destroyed. The MBH contains not only the photoperiodic timer, but also – and most remarkably, considering it is buried beneath the skin and bones of the avian skull – the photoreceptors that detect daylength. It has been known since the 1930s, when Jacques Benoit, a French physiologist, removed the eyes from mallard ducks and exposed the birds to long days, that the blind birds still had a photoperiodic response. Later work showed that this had to mean that the brain of birds contains photosensitive cells. Blind birds have no vision, but they can still sense daylength! On reflection, it is not such an odd idea. Light can penetrate deep into the body, as we have all discovered as children when we placed our hands over a torch under the bed-covers. It passes easily through the lightweight, translucent skull and brain of a bird (Foster *et al.*, 1985).

Almost 50 years after Benoit's studies, the photoperiodic response was triggered by implanting fine fibre optics into the bird brain to illuminate the MBH and simulate a local long day. These so-called 'deep brain photoreceptors' remain poorly characterised but are likely to be a group of light-sensitive neurones that sit near the ventricles within the MBH and detect dawn and dusk (Foster, 1998). The fact that neither birds nor mammals use retinal rods and cones as their primary means of detecting photoperiod emphasises the different sensory demands of 'dawn/dusk' detection and image detection. The former requires the measurement of light over a period to build up a reliable impression of the amount of light in the environment, and hence the duration of the day and night, whereas vision demands an instant 'snapshot' of light reflected from objects in the environment (Roenneberg & Foster, 1997).

Many of the clock genes found in the SCN of mammals are also found in the MBH of birds and, like mammals, these genes show 24-hour oscillations. The assumption is that there is a molecular clock in the MBH of birds that drives the photo-inducible phase (Figure 4.2). This is entirely consistent with an External Coincidence model of photoperiodism (Yasuo *et al.*, 2003).

Experimental evidence, originating from Brian Follett's group at the University of Bristol, showed that thyroid hormones are also crucial for a photoperiodic response in birds, and this work largely pre-dates the work in mammals. Removing the thyroid gland of seasonally breeding birds such as the Japanese quail (*Coturnix*) and European starling (*Sturnus*) blocked seasonal responses to photoperiod. Thyroxine ($T_4$) injections restored the photoperiodic response (Dawson *et al.*, 2001). A Japanese research team working on the quail showed a rise in $T_3$ levels in the hypothalamus during increasing daylengths and a decrease during short daylengths (Yoshimura *et al.*, 2003). Very recent studies in the Japanese quail support the hypothesis that PT-derived TSH acts locally within the MBH to elevate Dio2 (Nakao *et al.*, 2008) (Figure 4.11). The precise mechanisms by which the daylength information from the deep brain photoreceptors and the circadian clock are encoded into a signal that acts on the PT to produce TSH is still unknown, as is the means by which the TSH-driven production of $T_3$ by Dio2 then alters the activity of the GnRH neurones and other hypothalamic pathways.

Circadian clocks are used to measure the duration of the photoperiod in birds and mammals, and this is used to drive seasonal physiology. There is also evidence for the involvement of both External Coincidence (birds) and Internal Coincidence (sheep) photoperiodic timers. There is, however, very little information about how the still somewhat mysterious yearly circannual clocks are involved in the timing of annual events. These clocks were first studied in mammalian hibernation and bird migration (see Chapters 5 and 6), but they are now known to contribute to seasonal reproductive events. For example, if sheep are isolated from changes in environmental photoperiod and artificially maintained under a constant daily 12-hour light and 12-hour dark L:D cycle with no change in daylength, they will still show a circannual cycle in reproduction (Lincoln *et al.*, 2005). This suggests that although there is a circadian-based photoperiod timer there is also a separate circannual timer. In mammals, the generation of long-term endocrine cycles may depend on an interaction between a circadian-based, melatonin-dependent timer that drives the initial photoperiodic response and a

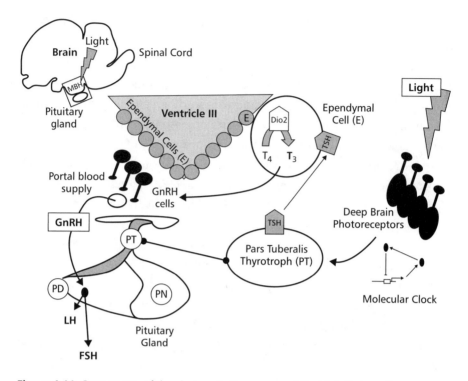

**Figure 4.11** Components of the photoperiodic response of the bird. Birds, unlike mammals, do not use pineal melatonin as a physiological representation of night length, although it is possible that melatonin may have some subtle modulatory effects on the photoperiodic response. The medial-basal hypothalamus (MBH) contains both the photoperiodic timer and the photoreceptors that detect daylength. Photoperiod is measured using a molecular clock, and long days stimulate cells (thyrotrophs) in the pars tuberalis of the anterior pituitary to release thyroid-stimulating hormone (TSH), which travels to the ependymal cells of the third ventricle of the hypothalamus. TSH stimulates Dio2, which converts thyroxine ($T_4$) into its active form, $T_3$. $T_3$ then stimulates the neurosecretory gonadotrophin-releasing hormone (GnRH) neurones. GnRH stimulates the pituitary gland to release the hormones LH (luteinising hormone) and FSH (follicle-stimulating hormone). These travel in the blood and reach the reproductive organs to regulate different aspects of reproductive physiology, including in males the production of testosterone from the testis, and in females oestrogen from the ovary.

non-circadian-based timer that drives circannual rhythmicity in long-lived species (Lincoln *et al.*, 2006).

Eberhard Gwinner demonstrated similar circannual rhythms in reproductive cycles of different bird species. The birds he studied differed in the conditions under which they expressed circannual rhythms. Long-distance migrants, which naturally experience a wide range of photoperiods during their annual journeys (Chapter 6), had particularly robust circannual rhythms. Studies of garden warblers indicated that circannual rhythms and photoperiod together affect reproductive timing (Gwinner, 1996b). When the birds were kept under a constant daylength, their response to a stimulatory photoperiod increased gradually, until, around the time of natural reproductive reactivation, the birds initiated circannual regrowth of their reproductive system. The existence of a circannual timer makes it likely that the circadian-based photoperiodic regulation of reproduction is only one aspect of the mechanisms that can be used to time reproductive effort across the seasons.

Quite where this circannual timer may be and how it works are almost entirely unknown. A recent proposal is that in sheep a circannual rhythm generator is centred in the pituitary gland. The idea is based on the interaction between melatonin-responsive cells in the PT that act to decode photoperiod, and lactotroph cells of the adjacent pars distalis that secrete prolactin. A mathematical model of this interaction produces, via a delayed negative feedback mechanism, a self-sustained, circannual rhythm in endocrine output (Macgregor & Lincoln, 2008).

Research over the past 100 years has told us much about the mechanisms that generate a photoperiodic response in birds and mammals. For most of this time these mechanisms have been thought to differ profoundly. It is only in the past few years that researchers have found that birds and mammals have much more in common than originally envisaged. In both groups, increasing daylengths promote TSH secretion from the pars tuberalis of the anterior pituitary. TSH then travels to ependymal cells of the ventral hypothalamus, which stimulate Dio2 enzyme activity, which converts $T_4$ to $T_3$. $T_3$ is then responsible for the regulation of pituitary hormones involved in the photoperiodic response. The final links in the photoperiodic chain seem broadly conserved between birds and mammals; the differences occur at the front of the chain, where daylength is encoded. Birds use deep brain photoreceptors and some form of circadian and circannual timer to regulate TSH production in the PT. By contrast, mammals have 'replaced' deep brain photoreceptors with specialised photosensitive ganglion cells within the eye and a melatonin signal that impinges directly on the PT to regulate TSH production. Quite why and when the birds and mammals diverged in this regard is now the subject of many PhD theses.

The story of the seasonal timing of reproduction in mammals and birds is complicated, but by unravelling the causal chain we are beginning to learn much more about reproductive health, which promises to have significant effects in the management of livestock and the number and timing of their progeny. For instance, two complete cycles of reproductive activity/inactivity in sheep can be induced in a 12-month period by doubling the daily rate of change in daylength (Notter, 2002). This effectively increases reproductive capacity by 100 per cent! The process is not that straightforward and depends on the species, but if global meat prices increase, farmers will no doubt be looking at ways of increasing supply.

# 5

# STAYING PUT IN THE COLD: HIBERNATION AND DIAPAUSE

A sad tale's best for winter: I have one
Of sprites and goblins.

WILLIAM SHAKESPEARE, THE WINTER'S TALE, ACT 2, SCENE 1

As summer shades into autumn in the higher latitudes, the shortening days and falling temperatures make for a very different environment. The onset of winter itself brings frost and a scarcity of food and sunlight. Deciduous trees cope by dropping all their leaves and shutting off photosynthesis. They dismantle their photosynthetic biochemistry and withdraw nutrients such as sugars, amino acids, nitrogen, phosphorus and potassium into their branches, trunks and roots. The loss of chlorophyll reveals other pigments, such as orange carotenes and yellow xanthophylls, which produce the spectacular colours of autumn in the temperate deciduous forests.

When the leaves are shed, an abscission zone forms where the leaf stalk meets the stem, cutting off the supply of water and nutrients. This leaf abscission is controlled by a complex system of hormones, responding to daylength, lower temperatures and light intensity (Raven *et al.*, 1999).

Many plants form storage organs below the ground while the rest of the plant dies. Unlike animals, very few plants store energy reserves as fats but more usually as starch. Starch binds water, which also helps prevent desiccation and reduces the freezing point of water. These plant storage organs provided the means for many animals, including our recent ancestors, to survive and feed in the northern winters (Thomashow, 1999).

Farther north, coniferous trees are the most prominent evergreen plants. Lack of water, which is locked up as ice above and below the ground, is their main problem. The most important and obvious adaptation to this is needle-like leaves, which help in greatly reducing the rate at which water is lost from the leaf surface. Evaporation is reduced even more by the presence of a thick, waxy cuticle and by locating the leaf pores (stomata) in deep sunken pits. But even these adaptations cannot sustain coniferous trees during the harshest winter conditions. The Arctic tree line is the most northern point where trees can grow. Beyond this point the extreme cold will freeze internal sap, and permafrost prevents trees from getting their roots deep enough into the soil for the necessary structural support. Where the Arctic tree line is drawn is heavily influenced by local variables, for instance the degree of slope and maritime influences such as ocean currents (Raven *et al.*, 1999).

Unlike plants, many animals respond to seasonal change by moving and migrating to a less harsh environment. However, a large number of animals stay where they are, relying on changes in their physiology and behaviour to enable them to survive the winter. In some species the adults die but leave offspring in an arrested stage of development to emerge in the spring (McNab, 2002). Yet other species hibernate or at the very least become dormant.

The winter cold is the primary killer for both plants and animals. Frost is particularly dangerous because living things are largely made up of water. Water is everywhere, both within the cells and in the spaces between them. If the water freezes and turns to ice, it expands and the ice crystals tear apart cell walls and plasma membranes. This is bad enough, but the loss of water also greatly increases the concentration of salts and organic molecules (solutes) in the extracellular compartments. At high concentrations, water is pulled from the intracellular fluids by osmosis, causing cellular collapse.

Cold winters pose a particular problem for amphibians. Frogs, toads and newts are generally small animals – although the Chinese giant salamander (*Andrias davidianus*) can grow to a metre in length – and they live in relatively small territories. Amphibians are also famously ectothermic (cold-blooded), so most of the time the body temperature is close to that of their surroundings. They avoid the lethal effects of the winter freeze in the higher latitudes by digging under a large object such as a rock, or deep into the soil to a level that does not become frozen, and rely on agents in their tissues that act as antifreeze to prevent damage to the cells. However, the farther north they are, the more drastic and spectacular is the physiological response to the intense cold. The American wood frog (*Rana sylvatica*) astounded Captain Francis Smith, who reported on it in his ship's log near the Canadian Arctic Circle in May 1747. With the onset of the shortening days of winter, the wood frog buries itself in the soil, often occupying a burrow made by another animal, and becomes immobile. As the temperature falls, the frog's toes begin to freeze. This triggers the conversion of glycogen stores in the liver to glucose, which enters the blood and from there it travels to all the intracellular and extracellular spaces of the body. The glucose acts as antifreeze, lowering the freezing point of water (a process called supercooling) and preventing the formation of ice crystals and the movement of water out of cells. Apart from safely freezing, the wood frog stops its heart beating and ceases to breathe, but as Captain Smith observed, 'Come spring, when the land thaws, so does his body. Within an hour or two the Frog will recover his Summer Activity, and leap as usual' (Stefansson & McCaskil, 1938). The Asian salamander (*Hynobius kyserlingi*) and the grey treefrog (*Hyla versicolor*) use glycerol rather than glucose as antifreeze, suggesting that the supercooling adaptation may have evolved independently in different amphibia.

Insects are also ectothermic animals, and cold-hardy species overwinter by similar basic strategies: freeze tolerance by withstanding the formation of a certain amount of internal ice by limiting the presence or location of ice in the body, and freeze avoidance by supercooling. Insects protect their body against the structural damage of ice crystals by restricting the formation of ice to extracellular spaces. This allows some insects to survive laboratory temperatures as low as $-50°C$. The key to this remarkable ability is ice-nucleating agents (INAs) that promote the freezing of fluids extracellularly rather than intracellularly. In addition to INAs, these freeze-tolerant insects use sugars for protection (Sinclair *et al.*, 1999).

Birds and mammals are endotherms (warm-blooded) and in theory at least can maintain a more or less constant internal temperature all year round. Mammals that remain active in winter maintain a core body temperature of around 37–38°C; for most birds it is a little higher, at around 40°C. In the Polar regions the environmental change is even greater, ranging between −60°C and +10°C. Birds and mammals use a menu of adaptations to cope with the extreme cold, including raising the metabolic rate to produce more heat to counterbalance heat loss; increasing the physical insulation between the body and the environment to reduce the rate of heat loss; altering the pattern of blood flow around the body to reduce blood flow, and hence heat loss, at the extremities and skin; laying down food stores to provide energy over the winter period; and adopting social behaviours that promote individual survival in such harsh conditions (Irving, 1966).

Many species in the colder regions escape from the extreme conditions by entering a state of 'physiological inactivity'. A broad range of terms have been used to describe these inactive states, including dormancy, torpor and hibernation. Dormancy is defined as a period in an organism's life cycle when growth, development and physical activity are almost completely suspended temporarily. Torpor can refer to a short-term state of decreased physiological activity, usually characterised by a reduced body temperature (hypothermia) and rate of metabolism. Many animals, including hummingbirds and bats, go through a daily torpor, showing normal body temperature and activity levels during the day and lowered body temperature and inactivity at night to conserve energy. In winter, torpor can last for months and all through the day and night. In seasonal torpor, its onset and finish are largely determined by temperature changes. Skunks (*Mephitis mephitis*) and badgers (*Meles meles*) may undergo periods of winter torpor as an energy-saving measure, but only during times of extremely cold weather.

Hibernation is somewhat different. It usually refers to an adaptive winter physiology of sustained inactivity and metabolic depression, characterised by a profound lowering of body temperature, breathing and metabolic rate. Body temperature falls, from around 38°C to as low as 1°C above ambient temperature. The heartbeat becomes slow and irregular. Breathing may be once every four to six minutes, and a hibernator's metabolic rate falls to as little as 1 per cent of its normal value. Entry into and exit from hibernation seems to be heavily dependent on a seasonal timer.

The terms obligate and facultative hibernators are sometimes used to distinguish between hibernation and winter torpor. The Arctic ground squirrel (*Spermophilus undulatus*) is an obligate hibernator. It enters hibernation around 5–12 October and emerges between 20 and 22 April, regardless of the weather, because it is programmed by a timer to do so. Skunks and badgers are facultative hibernators, entering torpor in response to very cold weather and poor food supply. The North American pocket mouse (*Perognathus californicus*) is another facultative hibernator.

While several temperate fish have seasonal temperature-induced torpor, only one species found so far hibernates, or at the very least comes close to it. The Antarctic 'cod' (*Notothenia coriiceps*) can enter a dormant state, similar to hibernation, that is not simply temperature driven. Antarctic fish have very low metabolic rates at the best of times, and this 'cod' has a heart rate of only ten beats a minute even in summer. In winter, it switches its ecological strategy from one that maximises feeding and growth to another that minimises the energetic cost of living during the long, Antarctic winter even though food is still available.

Describing the finding, Hamish Campbell of the University of Queensland, said (Campbell *et al.*, 2008):

> Fish are generally incapable of suppressing their metabolic rate independently of temperature. Therefore, winter dormancy in fish is typically directly proportional to decreasing water temperatures. The interesting thing about these Antarctic cod is that their metabolic rates are reduced in winter even though the seawater temperature does not decrease much. However, there are big seasonal changes in light levels, with 24 hour light during summer followed by months of winter darkness – so the decrease in light during winter may be driving the reduction in metabolic rates.

Although Aristotle believed that birds hibernate in holes in the ground, in truth only one bird species comes close. The Common Poorwill (*Phalaenoptilus nuttallii*) is a nocturnal bird that is found in much of North America and shows a marked degree of winter torpor. It spends the winter months inactive, slowing down its metabolic rate and dropping its body temperature. In periods of cold weather it may stay in torpor for several

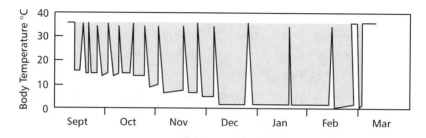

**Figure 5.1** Record of body temperature from September to March for a Richardson's ground squirrel. During much of the hibernation period, body temperature is close to 0°C, but during short episodes of arousal the body temperature rises to almost normal levels. The energy demands of arousal are considerable. Around 80 per cent of the energy used during hibernation is associated with the arousal periods (Pough *et al.*, 1996).

weeks. This behaviour was recorded in North Dakota in 1804 by Meriwether Lewis during the famous Lewis and Clark Expedition overland to the Pacific coast.

Although mammals seem to be 'in suspended animation' while hibernating, it is not a passive process but one of active management. Body temperature is maintained a little above the ambient surroundings by lowering the set-point for thermoregulation, but there are additional physiological mechanisms that act on body temperature. This active management is seen through the study of arousal periods during hibernation whose frequency and length vary widely between species, individuals and the time of year. The precise function of this periodic arousal is not clear, but it must be important because the energy expended in a single arousal lasting only a few hours can equal that used in ten days of hibernation. The relatively brief periods of arousal in the Richardson's ground squirrel (*Spermophilus richardsonii*) constitute approximately 80 per cent of all the energy used during the entire hibernation episode (Figure 5.1).

It has been assumed that arousals allow an individual to feed on food stores in the hibernation den. Chipmunks do this, but many species do not and only metabolise stored body fat during hibernation. It seems most likely that appetite must be switched off or else the animal will be hungry and need to search for food, obviating the point of hibernation. But if the animal does not eat, it will lose weight; although that is normally a prompt to look for food, during the hibernation this critical physiological control variable is also reset at a lower level.

Even if the body weight is allowed to drop, hibernating animals still require internal energy reserves. Hibernators typically feed intensively before winter to build up their fat stores. Some, such as the edible dormouse (*Glis glis*), fill up on carbohydrate- and lipid-rich foods such as seeds in the run-up to hibernation. Many bats build up fat reserves in the autumn that represent as much as a third of their total mass. Much of this fat is stored as brown adipose tissue (BAT). This tissue gets its name not because it was discovered in bats but rather because of its dark colour, which comes from the dense plexus of blood capillaries that permeate the tissue. Brown adipose tissue is rich in mitochondria with special properties that enable it to oxidise fatty acids and/or glucose to produce heat very rapidly. Deposits of brown adipose tissue are found around some internal organs and between the shoulder-blades of hibernators. Their function is to generate body heat quickly, especially during periods of arousal. Getting the right balance between building a sufficient reserve and conserving it appropriately during hibernation is tricky. Sometimes these fat reserves run out. Belding's ground squirrels (*Spermophilus beldingi*) live at high altitude in Tioga Pass, California, and may hibernate for seven to eight months of the year. A staggering 60–70 per cent of all juveniles hibernating for the first time, and one-third of adult animals, are estimated to die during hibernation. Most of the deaths are thought to result from the burning-up of food reserves (Pough *et al.*, 1996).

What goes in – no matter its source – must come out, and most hibernating species will urinate and defecate, move about and change their position during arousal, suggesting that it provides an opportunity for various essential physiological processes such as the removal of waste and the repair and maintenance of cellular processes. During hibernation, the kidney has minimal blood flow but even though the metabolic rate of a hibernating animal is very low, nitrogen waste accumulates in the tissues. Because it cannot be removed by the production of urine, this nitrogen waste will eventually become toxic. Periodic arousal from hibernation is associated with an increase in blood pressure, kidney function and restored urine production.

Intensive protein synthesis during arousal supports the idea that arousal may be associated with cellular repair and maintenance. Paradoxically, given the common historical view that hibernation is a 'deep sleep', the drop in body temperature in hibernators is also associated with the loss of the brain activity associated with sleep. Counter-intuitively, these observations have led some researchers to hypothesise that animals may actually become sleep deprived during hibernation and that they arouse periodically so as to catch up on their sleep (Heller & Ruby, 2004).

In the popular imagination, bears are the prime example of hibernators, but there has been much discussion about whether bears do in fact truly hibernate. The consensus is that they do not, on the basis that they do not show the profound change in their physiology that is the mark of true hibernators. Black bears, for example, show classic winter torpor. The heart rate decreases from 40–70 beats per minute to 8–12 beats per minute; metabolism falls but by only 50 per cent, and body temperature dips only slightly, from about 38°C to 31–34°C. The bears' thick winter fur coats and their low surface area to volume ratio compared with smaller hibernators reduce heat loss, so they can stay warmer. They can, however, stay in this torpid state without eating, drinking, urinating or defecating.

More drastic metabolic shut-down in female bears is probably precluded because winter torpor is associated with cub development and birth. Mating in black bears occurs in early summer, before most berries and nuts ripen. Implantation of the fertilised eggs is, however, delayed until November, so that birth in the dens occurs in January. The newborn cubs are tiny, weighing only 200–450 grams each, which is approximately 1/250 of the mother's weight. The cubs are fed on milk, and although the mother is in winter torpor, she is still alert to her cubs' needs, responding to vocal demands for warmth, comfort and suckling. By delaying implantation and timing birth in the den to midwinter, the cubs spend less time confined in the den than if they had been born in late autumn or early winter.

The length and depth of the winter torpor matches the regional norms of food availability. In the northern portion of the black bear range, where abundant, high-quality food is available only from May to August, torpor can be sustained for over seven months. If food has been scarce in the late summer, and if fat reserves are small, some bears show a deeper level of torpor and they can be handled for several minutes before they wake up. It may be no more than a passing thought, but perhaps our ancestors took advantage of this vulnerability and hence played a part in the extinction of the giant Pleistocene cave bear some 29,000 years ago. By contrast, in southern states where food is available all year round, some black bears do not show torpor at all, and those that do are easily aroused. In black bears, the timing of entry into and exit from torpor are largely driven by environmental variables rather than an internal timer.

Numerous animals spend many months of the year underground in constant darkness and in a lowered metabolic state. There must be a signal that switches them out of this state of inactivity, but what is it? The superficial explanation is that they respond to some environmental factor such as temperature. Although this may well be true in some species, as far back as 1837 Arnold Berthold speculated that there might be some internal annual timer. More than a century later, in the 1950s and early 1960s, the Canadians Ted Pengelley and Ken Fisher provided unambiguous evidence for the existence of such a circannual clock in the golden-mantled ground squirrel (*Spermophilus lateralis*). Their initial experiments maintained wild-caught ground squirrels in a constant LD 12 : 12 cycle and at a constant temperature in the laboratory. Despite the constant cycle, the animals continued to hibernate about once a year and each hibernation episode was preceded by a marked increase in body weight and food consumption. In the first year of the experiment, animals entered hibernation in late October, but in the second year the start of hibernation had advanced to mid-August, and in the third year it occurred in early April. The behaviour resembled a free-running circannual rhythm of a period of approximately 10–11 months. Even golden-mantled ground squirrels that had been born in captivity and had never experienced a natural photoperiod still had a circannual rhythm (Pengelley & Fisher, 1966). Later experiments in the 1970s studied groups of squirrels for over 47 months in constant darkness (DD), constant light (LL) or LD 12 : 12 conditions. In all groups, hibernation continued about once a year, with a period usually shorter than 12 months irrespective of the lighting conditions (Figure 5.2).

Similar circannual hibernation rhythms have been found in other species of ground squirrel and in chipmunks, woodchucks, the pocket mouse and the dormouse. Circannual rhythms not only seem to play an important

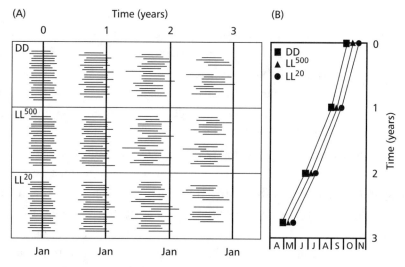

**Figure 5.2** An illustration of the circannual rhythms of hibernation in three groups of golden-mantled ground squirrels kept for nearly 4 years at 3°C under different photoperiods. Abbreviations: DD, constant dark; LL$^{500}$, constant light of 500 lux; LL$^{20}$, constant light of 20 lux. (A) Horizontal black bars represent hibernation periods of individual squirrels under these conditions. (B) The mean dates (November (N) to April (A)) at which the animals in the three groups entered hibernation in successive years of the experiment. The results show the persistence of a free-running circannual rhythm of hibernation under a variety of constant conditions (Pengelley et al., 1976).

part in the seasonal breeding of large mammals such as the sheep (see Chapter 4), but they also seem to be a critical feature of hibernation. But studying these rhythms requires enormous patience. It took Pengelley and Fisher years to collect their data from the golden-mantled ground squirrel. There remain a few groups around the world, however, brave enough to work in this demanding area of physiology. Noriaki Kondo and colleagues based in Tokyo have identified a hibernation-specific protein (HP) complex in the blood of hibernating chipmunks. They have shown that if chipmunks are kept in constant cold and darkness, HP is regulated by an individual free-running circannual rhythm that correlates with hibernation behaviour. The level of HP complex in the brain increases at the same time as the onset of hibernation, and blocking the activity of HP within the brain caused a decrease in the duration of hibernation. The suggestion is that HP, under the regulation of a circannual timer, carries hormonal signals that are essential for hibernation (Kondo *et al.*, 2006).

In the wild, circannual clocks run precisely at 12 months, but in constant conditions the rhythms have a period that is usually shorter than 12 months (Figure 5.2). This suggests that some signal (zeitgeber) in the environment entrains the clock. The most obvious zeitgebers are photoperiod, temperature and social cues. There is good evidence that photoperiod can act as a zeitgeber for circannual rhythms of reproduction and other seasonal behaviours (see Chapter 4). Photoperiodic entrainment for hibernation rhythms has been more difficult to establish. Changing the photoperiod systematically or shifting the photoperiod by six months causes some change in timing in ground squirrels, but the effects produced can only be described as inconclusive. It is probable that photoperiod acts as a zeitgeber at some level, but because the circannual rhythms of ground squirrels show considerable variation between individuals and because experiments have not studied animals over a sufficiently long period, this has been difficult to demonstrate (Gwinner, 1986).

Temperature can affect the period of circannual hibernation rhythms and may also contribute to their entrainment – perhaps acting synergistically with photoperiod. Pengelley and Fisher again led the way with studies on the golden-mantled ground squirrel. Squirrels were held on a LD 12 : 12 cycle at 35°C for different periods before being transferred to 0°C. The drop in temperature triggered hibernation, but in those squirrels exposed to 35°C for nine months or more before being transferred to 0°C the next hibernation period was shifted by about half a year. Mrosovsky has also provided strong evidence for temperature entrainment. He kept golden-mantled ground squirrels for a period of three years under a LD 12 : 12 cycle at 21°C. In these conditions the circannual rhythm of body weight loss and gain free-ran with a period shorter than 12 months. In parallel, an experimental group was exposed for nine months to −3°C at the end of the first year and was then transferred back to 21°C. This nine-month cold shock in the experimental group delayed the circannual rhythm of body-weight change by 4.5 months. These and similar studies on the European hamster and the common dormouse suggest that temperature might act along with photoperiod as an important zeitgeber in entraining hibernation rhythms (Gwinner, 1967).

There is very little that we understand about these circannual rhythms. The initial thinking was that circannual rhythms were the product of some mechanism that counted the circadian days encoded within the suprachiasmatic nuclei (SCN). This now seems highly unlikely because destruction of the SCN in golden-mantled ground squirrels does not block the expression of circannual rhythms of hibernation and body-weight gain (Ruby *et al.*, 1998). Further, lesions in different places of the hypothalamus and surrounding brain regions, and even removal of the pineal gland, have remarkably little effect on hibernation rhythms (Hiebert *et al.*, 2000).

Beyond Gerald Lincoln's idea that the circannual timer of sheep might reside within the PT of the pituitary gland (see Chapter 4) (Macgregor and Lincoln, 2008), we have little idea as to the anatomical location of the circannual clock in mammals, or even whether there is a distinct structure, let alone its precise mechanism. This may seem extraordinary in this age of biological triumphalism, but the lack of information reflects both the general difficulty of trying to associate specific regions of the brain with a specific physiological function and the truly enormous problem of studying a behaviour that occurs once a year.

A conspicuous adaptation in mammals that stay put in northern high latitudes is the thicker winter coat. The Arctic lemming can raise its metabolism to cope with severe cold, but it is very demanding of energy to use for extended periods of time unless there is abundant food; instead, most severe weather adaptations are associated with insulation and the prevention of heat loss. The insulation is provided primarily by feathers or fur, both of which are remarkable insulators when dry. Because static air is a very poor conductor of heat, air trapped under feathers and fur greatly reduces the heat loss between the animal and its environment.

Bird plumage and mammal fur change with the approach of winter. This preparation occurs in anticipation of the severe weather and is not triggered by it. The shortening photoperiod of autumn drives the moult, during which time birds change all or many of their feathers and mammals grow a thicker winter coat. In many species moult, too, shows circannual cycles. There are some birds that under a constant daylength will show circannual cycles of moult but not reproduction, illustrating once again that both circadian and circannual timers regulate seasonal physiology (Gwinner,

1996b). Birds have two sorts of feather: waterproof contour feathers, which form the exterior surface; and down, which lies beneath the contour feathers and traps much of the air. In winter, the house sparrow (*Passer domesticus*) has 11.5 per cent more down feathers than in summer. Each feather has an individual muscle that enables it to be raised away from the body. This enables the bird to regulate heat loss by altering the thickness of the air layer trapped among the feathers.

Mammals also possess two types of fur that make up the coat: fine hairs close to the skin to trap air, and coarser surface hairs to form a semi-impermeable surface. The longer the hairs, the better the insulation. Arctic mammals tend to have more hairs per unit area of skin than other mammals, and this gives them a slight insulation advantage. The insulating properties of both feathers and fur are abolished if the trapped air is replaced by water – which has approximately 25 times the heat capacity of air. So the black skin of the polar bear is entirely covered (with the exception of the nose and the foot pads) by layers of oily, water-repellent fur. When a bird preens, oil from the preen gland waterproofs the feathers, and this, along with the structural properties of the contour feathers, makes them a more effective insulator than fur (Davenport, 1992).

At the beginning of winter, birds and mammals often change social behaviours. Birds that have spent the spring and summer as breeding pairs and have been highly aggressive towards other individuals become gregarious and form large, cohesive flocks. As well as protection in numbers, feeding is more efficient because individuals can observe the feeding activities of flock-members. Another benefit of being social is that it can reduce heat loss. Pallid bats (*Antrozous pallidus*) expend less energy when they roost huddled together than they do if they roost alone (Irving, 1966).

Honeybees also huddle in winter. The answer to the question, 'where do honeybees go in winter?' is that they stay in their hives bundled close together for warmth and living off the honey stores they have made. Some insects migrate, and in autumn monarch butterflies fly thousands of kilometres to overwinter. But in non-equatorial latitudes, most insects restrict reproduction and growth to the warmer months of spring and summer, and the colder winter months are spent in a state of diapause. Insect diapause is a state of greatly reduced metabolic activity and in most cases is required for continued development. In the early 1920s, Simon Marcovitch (a Russian émigré who worked in the same building as Garner and Allard and later became a professor of entomology at the University of Tennessee) showed that seasonal forms of aphids are controlled by photoperiod (Marcovitch, 1924). This is usually triggered by the shortening daylengths of autumn, and insect diapause has been one of the most important models in the development of our understanding of photoperiodic phenomena.

All insects start as eggs and end as adults, but the path between splits broadly in two. Flies, butterflies and beetles are among those known as holometabolous insects. This group has a four-stage life cycle consisting of egg → larva → pupa → adult. Inside the pupa (chrysalis) the insect undergoes a complete transformation from larva to adult, and holometabolous insects exhibit what is known as 'complete' metamorphosis – the adult looks nothing like the larva or pupa. In contrast, hemimetabolous insects, which include cockroaches, grasshoppers, crickets, stick insects and aphids, have a three-stage developmental cycle of egg → nymph → adult. Hemimetabolous insects hatch as nymphs, which in most cases resemble a wingless miniature version of the adult (Figure 5.3). Unlike the holometabolous insects they lack a distinct larval/pupal stage. Furthermore, hemimetabolous insects develop their wings through their various nymph stages, which are well known to crossword solvers as instars, whereas holometabolous insect wings are never seen until the adult insect emerges (Gullan & Cranston, 2004). To complicate matters even further, holometabulous insects may go through several larval growth stages, also known as instars.

Different insect species can arrest development as eggs/embryos, larvae, pupae or adults. Mosquitoes (*Aedes*) enter diapause as embryos within the egg; some wasps (*Nasonia*) as larvae; flesh flies (*Sarcophaga*) as pupae; and bugs (*Pyrrhocoris*) as adults (Saunders *et al.*, 2002). The term diapause was first used in 1893 by the Texan biologist William Wheeler to describe this arrested stage. A modern, and much cited, definition is by Maurice Tauber and colleagues at Cornell University (Tauber *et al.*, 1986):

A neurohormonally mediated, dynamic state of low metabolic activity. Associated with this is reduced morphogenesis, increased resistance to environmental extremes, and altered or reduced behavioral activity. Diapause occurs during a

| Holometabolous Insects | Hemimetabolous Insects |
|---|---|
| 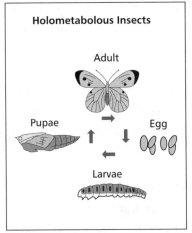 | 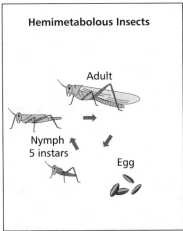 |

**Figure 5.3** Holometabolous insects such as the large white butterfly (*Pieris brassicae*) lay their eggs on the underside of leaves, for example cabbage or nasturtium. Eggs hatch into a larva (caterpillar) and after several moults (instars) leading to ever larger larvae the cuticle begins to darken and harden to form a pupa. During winter the pupa is in a state of diapause, from which the adult emerges in the spring. Hemimetabolous insects such as the desert locust (*Schistocerca gregaria*) lack a larval stage. Instead, a nymph emerges from the egg and undergoes a series of moults to produce instars (called 'hoppers'), which look increasingly like the adult. When they develop wings, they fly to a new area, where they feed, mate and reproduce. Adults undergo diapause in late summer and autumn and remain sexually immature until the following spring.

genetically determined stage(s) of metamorphosis, and its full expression develops in a species-specific manner, usually in response to a number of environmental stimuli that precede unfavourable conditions. Once diapause has begun, metabolic activity is suppressed even if conditions favourable for development prevail.

Diapause is not only about surviving a specific period of seasonal change but also about the synchronisation of entire life cycles to seasonal change in different environments. Hugh Danks of the Canadian Museum of Nature has explained this important idea succinctly (Danks, 2005):

Each individual follows a particular life-cycle pathway by making successive developmental decisions, such as whether or not to enter diapause, whether or not to become quiescent, and whether to develop slowly or rapidly. Many alternative pathways are possible when there are multiple decision points through the life cycle; each individual can then optimize its chance of survival by adjusting the duration of immature development, the time of metamorphosis to the adult, or the timing of reproduction according to its genetic programme and especially in response to ongoing environmental information. Of course, the response to a given environment can change through the life cycle, so that short photoperiods might have one effect in early instars and the opposite effect in later instars.

A key point about diapause in insects is that it is a process whose timing can be determined at one point in the life cycle and manifested in another. But no matter the stage, the critical event in diapause regulation is a block on the secretion of at least one neuropeptide from the brain. At low or even no levels of secretion, development is effectively 'stuck'.

Insect development relies heavily on three hormones (Figure 5.4). Pro-thoracicotropic hormone (PTTH) is produced in neurosecretory cells in the pars lateralis of the brain but stored and released from an organ called the corpus cardiacum. PTTH stimulates the prothoracic glands located behind the head to produce steroid hormones called ecdysones. Forms of ecdysone stimulate the moult and also promote adult development from the pupa. Juvenile hormone (JH) is synthesised and released from the corpus allatum. Its action is complex but, essentially, high concentrations of JH (with ecdysone) stimulate the production of another larval stage and so in effect delay the formation of a pupal stage in holometabolous insects. When JH falls below a critical concentration and/or target tissues become less responsive, a pupa can form, a development stimulated by ecdysone alone (Figure 5.4) (Gullan & Cranston, 2004).

Adult diapause is brought about by a shut-down in JH. This hormone is usually associated with insect development, but despite its name JH is present in adult insects and is involved in regulating aspects of sexual maturation, such as egg development in females. Reduced circulating levels of JH result in reproductive diapause. In the Colorado potato beetle (*Leptinotarsa decemlineata*), which can devastate potato crops, adult diapause is induced by short daylengths and broken by the increasing daylengths of spring.

Carroll Williams and his co-workers at Harvard University carried out

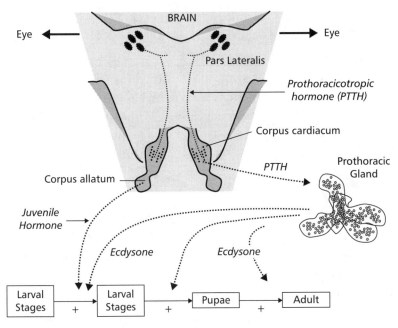

**Figure 5.4** View of the top of an insect brain (dorsal view). Three main hormones are involved in insect development. (1) Prothoracicotropic hormone (PTTH) is produced in neurosecretory cells in the pars lateralis of the brain and released from the corpus cardiacum. PTTH stimulates the prothoracic gland to produce the hormone ecdysone. (2) Ecdysone is secreted from the prothoracic glands and stimulates moult and the development of the adult. (3) Juvenile hormone (JH) from the corpus allatum promotes the production of another larval stage (instar) and so delays the formation of a pupal stage. At low levels of JH, pupal development occurs. In adults JH mediates reproductive activity such as egg development.

the classic work on pupal diapause studying the giant silkmoth (*Hyalophora cecropia*). In this species, pupal diapause occurs in the late summer and is triggered by the failure to release PTTH by the caterpillar. This in turn results in lowered levels of ecdysone from the prothoracic gland. Silkmoths become arrested as pupae, emerging as adults in May or early June. The adults live only about two weeks and cannot feed. Their sole purpose is to mate and produce eggs (Saunders *et al.*, 2002).

The photoperiodic signal for diapause is often detected at an early stage of development during a 'sensitive period' (Table 5.1). Pupal diapause in the fly *Sarcophaga* is triggered when the embryos are developing in the egg within the mother and in early larval development. It can be even earlier! In the wasp *Nasonia* and fly *Calliphora*, the photoperiod is detected by the mother and amazingly conveyed to the undifferentiated egg. How this happens we do not know, but it seems that during the sensitive period a certain number of inductive photoperiods need to be 'added up' before a threshold is reached and the diapause switch is thrown. Although daylength is the key environmental factor, moisture, diet and temperature can all influence the precise timing of diapause in different species. Generally, the colder the weather, the higher the incidence of diapause within the population.

| SPECIES | DIAPAUSE STAGE | SENSITIVE PERIOD |
| --- | --- | --- |
| *Bombyx mori* (moth) | Egg | Maternal |
| *Nasonia vitripennis* (wasp) | Larva | Maternal |
| *Sarcophaga argyrostoma* (fly) | Pupa | Embryo |
| *Pyrrhocoris apterus* (bug) | Adult | Nymph |
| *Calliphora vicina* (fly) | Larva | Maternal |

**Table 5.1** This table shows the stage of the life cycle at which each of these species enters a state of diapause and the stage in the life cycle at which the insect is sensitive to the photoperiodic information that triggers diapause.

Many insects use specialised light-detecting systems rather than their sophisticated compound eyes and ocelli to detect the photoperiod light signal. This is one of the many important findings that are largely due to Tony Lees. One of the early pioneers in insect photoperiodism (Saunders, 2005), his interest was sparked at Cambridge when he began work on the orchard pest, the red spider mite (*Metatetranychus ulmi*). He left some colonies of mites in a glass container on the laboratory windowsill and during autumn he noticed that they began, apparently spontaneously, to produce their overwintering red diapause eggs. He deduced that this must be a response to the shortening daylengths, because the temperature within the laboratory had remained relatively constant. He later moved on to study the photoperiodic responses of the vetch aphid (*Megoura viciae*), which he studied while at Imperial College, London (Jim Hardie, personal communication).

The life cycle of this aphid is more than a little complicated, and the

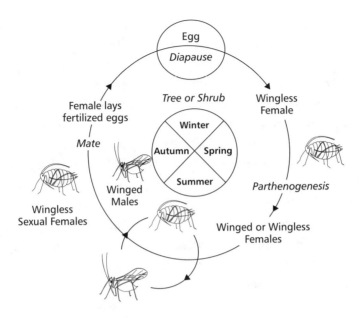

**Figure 5.5** The life cycle of the vetch aphid (*Megoura viciae*). The eggs survive the winter in a state of diapause. In spring the eggs hatch and give rise to wingless females, which in turn produce daughters by parthenogenesis (the production of offspring without the involvement of sex). These parthenogenic daughters are called 'virginoparae'. During the late spring and summer months, females produce either winged or wingless forms depending on crowding. Low population densities promote the production of wingless females, whereas high densities promote the production of winged adults. These winged females fly away to colonise new plants. In the autumn, when the days begin to get shorter, the virginoparae produce both winged males and wingless sexual females. After mating, the sexual females lay hard-shelled eggs, which overwinter in a state of diapause in the leaf litter (Hardie & Vaz Nunes, 2001).

details are shown in Figure 5.5. In essence, during the long spring and summer days, males are redundant. The adult female grows and develops eggs with neither contact nor help from a male, a process of asexual reproduction known as parthenogenesis. The eggs develop into females and they emerge from the mother as live births, a process known as vivipary. With the shortening days of autumn, females switch production to sexually mature winged male and female forms, which then mate to produce eggs that enter diapause over the winter months (Hardie & Vaz Nunes, 2001).

Lees used this complex life cycle as his diagnostic tool to hunt for the photoreceptors detecting dawn and dusk in aphids. Remarkably, given the equipment available in the 1960s and 1970s and the tiny size of the aphid, he was able to use fine optic fibres to illuminate specific regions of the insect's body and brain. The aphids were kept in a room on a photoperiod of LD 14 : 10 to mimic the photoperiod of early autumn. In addition, via the optic fibre to their brains, he provided the aphids with a two-hour supplementary light exposure each day. His logic was that if the two-hour light treatment was detected and effectively added to the background photoperiod, the aphids would be exposed to a photoperiod of LD 16 : 8 and this would be interpreted by the aphid as summer. Under these conditions the aphids would produce parthenogenetic daughters. However, if the two-hour light supplement was not detected, the aphids would translate the LD 14 : 10 photoperiod as early autumn and this would trigger the production of egg-laying sexual males and females. The most effective placement of the fibre optic for maintaining parthenogenesis was the dorsal midline of the head. If the light was directed at the eyes, or any other part of the body, 'autumn' daylengths stimulated egg-laying sexual forms. It was a clear result. The light was being perceived exclusively by photoreceptors within the brain (Lees, 1964).

It is not only aphids that have brain photoreceptors. Larval diapause in blow flies is induced by maternally detected short days, and if the maternal visual system is removed, larval diapause can still be induced by short days (Saunders & Cymborowski, 1996). Conclusive proof that the insect brain contains both the photoreceptors and the photoperiodic clock came from remarkable experiments on the isolated brain of the tobacco hornworm moth (*Manduca sexta*). The brain and associated neurosecretory structures (Figure 5.4) were removed from a larval moth on a short-day (diapause inductive) LD regime and maintained in culture medium. The brain was then exposed to three long-day cycles in the dish and implanted into a short-day larva from which the brain had been removed. After implantation into the new host, diapause was not triggered. The isolated long-day brain had redirected the larvae along a non-diapause line of development (Bowen *et al.*, 1984). Because nothing in this area of work is simple, the evidence that the photoreceptors are in the insect brain does not mean the compound eyes are not required for the detection of photoperiod, nor that they have no role. It seems that some insects do use their compound eyes along with brain photoreceptors for photoperiodic detection (Shiga & Numata, 2007).

The brain photoreceptors use a photopigment based on opsin and vitamin A that is maximally sensitive in the blue (450–470 nm) part of the spectrum (Shimizu *et al.*, 2001). Insect brain photoreceptors share the same basic opsin/vitamin A biochemistry as other animal photopigments and have maximum sensitivities in the same 'blue-light' part of the spectrum as the brain photoreceptors of birds (Foster *et al.*, 1985) and the melanopsin-based photosensitive retinal ganglion cells (pRGCs) (Hattar *et al.*, 2003).

Establishing the location and structure of the insect photoreceptors and the basic biochemistry has been a monumental piece of work, but it is a precursor to understanding the nature of the photoperiodic timer. The questions, as with plants, mammals and birds, have focused on whether insects use an hourglass or a circadian timer, and how it works. Many insects have been exposed to Nanda–Hamner protocols and Bünsow (night interruption) protocols (see Appendix I) to explore whether a circadian timer prevailed. In general, insects have a circadian timer as shown schematically in Figure 5.6, which presents the experimental data supporting the presence of such a circadian timer.

As we discussed in Chapter 3, Erwin Bünning had proposed that, in plants, light interacted with the circadian system in two ways: namely the entrainment of the rhythm to dawn and dusk, and the illumination of a light-sensitive phase of the cycle. As the daylength changed with season, this photosensitive phase of the cycle was exposed to light, leading to long-day responses. This could be put the other way round: rather than a particular phase being sensitive to light, it was light's absence that was critical in inducing photoperiod-driven change.

Detailed studies of various protocols revealed that the induction of diapause was indeed dark-sensitive and was inhibited by light falling at a particular time after dawn. This observation is schematised in Figure 5.7.

Although a circadian timer has been strongly implicated in generating a photoperiodic response in many insects, a significant number of experiments have shown 'negative' responses to Nanda–Hamner protocols. Working with his vetch aphids, Tony Lees found that when the 'night' exceeded a critical duration, a diapause response was triggered and that there was no circadian modulation of the response (Lees, 1966). This led Lees to conclude that there had to be an hourglass timer at work (Figure 5.8).

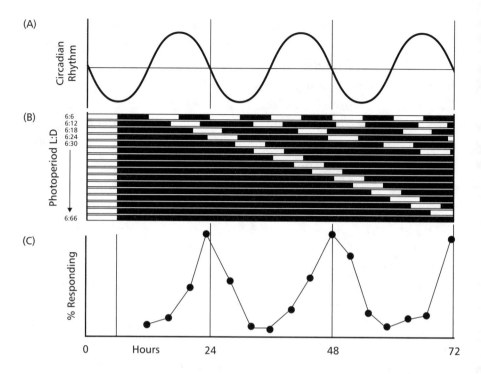

**Figure 5.6** (A) Depiction of a 24-hour circadian rhythm that drives a 24-hour rhythm in light sensitivity. (B) The Nanda–Hamner protocol in this case begins with 6 hours of light coupled to an increasing number of dark hours to give overall period (*T*) ranging from 12 to 72 hours. (C) Illustration of a 'positive response'. If photoperiodic induction occurs at or near 24 hours, or multiples of 24 hours, then an underlying circadian rhythm in sensitivity is implicated. Such responses have been described in flies (*Calliphora* and *Sarcophaga*) and the wasp *Nasonia* maintained at temperatures around 20–22°C.

To complicate matters even further, David Saunders at Edinburgh University, who is one of the doyens of insect timing behaviour, has found that positive and negative responses can be obtained in the same species depending on environmental conditions such as temperature, diet or even latitude. For example, in the flesh fly, hourglass-like responses can be observed at low temperatures (about 16°C) and circadian-like responses are observed at higher temperatures (about 22°C) (Saunders *et al.*, 2002).

So it would seem that in some species the timer is circadian and in others an hourglass, and in some cases it could be one or the other depending on the weather. This is not altogether satisfactory as an explanation, and this

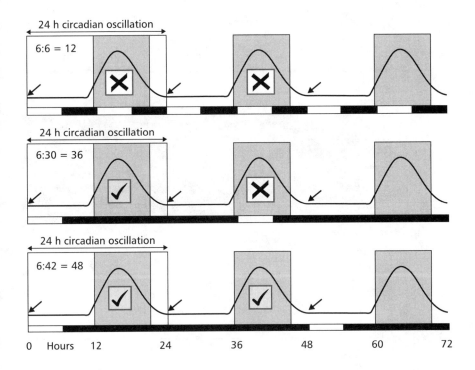

**Figure 5.7** Here we have assumed that there is a 24-hour circadian-driven rhythm (the arrow indicates the start of the rhythm) of 'dark induction' (grey box) such that if darkness falls between 12 and 20 hours after dawn, diapause will be induced. If the insect is exposed to light during this period, diapause will be inhibited. Under a Nanda–Hamner protocol of 6 hours of light and 6 hours of darkness (6:6, as shown by the white and black bars under the graph; period length ($T$) = 12 hours), light will always fall on the dark inductive phase and there will be no diapause; under a photoperiod of 6:30 ($T$ = 36h), the circadian-driven dark inductive phase will be alternately stimulated and then inhibited, again resulting in no photoperiodic induction of diapause. However, under a 6:42 regime ($T$ = 48 hours), the 24-hour rhythm of dark induction will always fall (as a multiple of 24 hours) during the dark portion of the photoperiod, and diapause will be induced.

controversy as to whether photoperiodic timing is circadian or hourglass in insects has been rumbling on for nearly 50 years. The mild-mannered Bünning offered the wise observation, 'nature has certainly more ways than one to solve such problems' (Saunders *et al.*, 2002), and it could be that different species have evolved with different solutions. Biology is untidy.

Influenced by his findings in the flesh flies, David Saunders and colleagues have argued that it is unlikely that an insect will switch from one

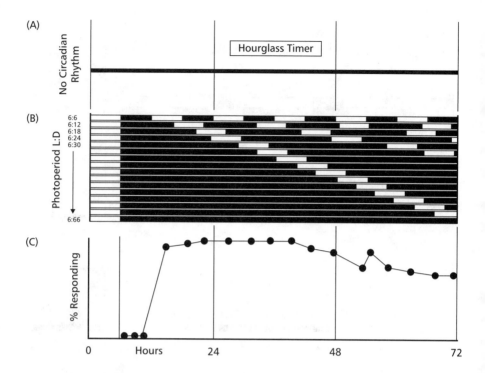

**Figure 5.8** (A) Depiction of a non-circadian hourglass timer: there would be no 24-hour gating of any responses. (B) The Nanda–Hamner protocol uses a 6-hour light period coupled to an increasing number of dark hours to give an overall period (*T*) ranging from 12 to 72 hours. This protocol is the same as that illustrated in Figure 5.6. (C) Illustration of a 'negative response'. Photoperiodic induction is triggered when a critical night length is reached and then reaches a plateau with no apparent circadian (24-hour) gating of the response. Such responses have suggested the involvement of an hourglass timer rather than a circadian timer. In this illustration (based on studies in aphids), when the night length exceeded 12–14 hours (resembling early autumn) sexual forms of aphids will be produced, ultimately leading to the production of overwintering eggs (see Figure 5.5).

form of timer (circadian to hourglass and back) with a change in temperature, latitude or diet. He has proposed that the variable results arise instead from a single mechanism that expresses itself differently under different conditions.

The broad explanation depends on the idea that a master clock or pacemaker is coupled to slave oscillators. The photoperiodic response is not generated by the master pacemaker but by the slave oscillators, which in turn drive a rhythm of photosensitivity (Figure 5.9). Normally the slave oscillators (coupled to the pacemaker) drive a circadian rhythm of dark induction such that if darkness exceeds a certain critical period (for example the increasing night lengths of autumn), diapause will be triggered. If the slave oscillators are very weakly coupled to the master pacemaker, they rapidly uncouple and damp out under conditions such as darkness or low temperature. The arrival of light will restore coupling between the master pacemaker and slave oscillators and thus the reappearance of a rhythm of dark induction. As soon as the critical night length has been exceeded, a diapause response will be triggered. So a heavily damped circadian oscillator will to all intents and purposes seem to be an hourglass timer!

Not everyone agrees that the answer has been found, and for the lay reader there is the obvious question: does it matter? It does if we are to advance our understanding of seasonal timing in insects, and this is important once we recognise how vital insects are in agriculture and how much is spent on chemical agents that are used to control them. If we knew how it worked, we might be able to design methods of intervening in the process.

So for the time being, on the assumption that insect diapause is based on a circadian oscillator, the next question is how this arrangement might work at the molecular level in anticipating seasonal change. The usual place to start would be to reach for the fruit fly (*Drosophila melanogaster*), because this animal has provided much of our conceptual understanding of the molecular basis of the circadian clock in animals in general. No other animal has been as well characterised. Frustratingly, *Drosophila melanogaster* has a very weak diapause response to photoperiod. As a result, the genetic and molecular analysis of the insect photoperiodic timer lags far behind its physiological and behavioural analysis and we have little idea of the molecular clockwork of those insects that show a robust diapause response to the changing photoperiod (Tauber & Kyriacou, 2001).

Although the photoperiodic response of *Drosophila melanogaster* is fairly weak in the laboratory at least, it is not altogether absent. Female flies exposed to short days and low temperatures enter an ovarian reproductive diapause in which adult females turn off egg production. In the higher latitudes of both hemispheres, individuals in the wild survive for several months through the winter and then resume egg-laying in the spring. The lack of

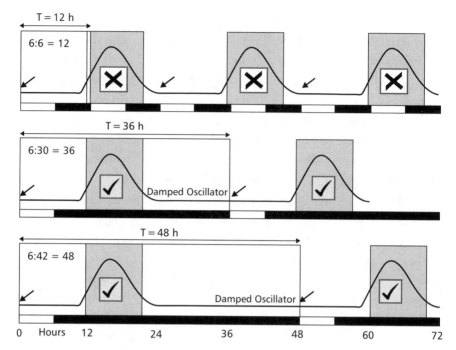

**Figure 5.9** Illustration of how a rapidly damping slave oscillator can give rise to 'negative' result in a Nanda–Hamner protocol (compare with Figure 5.7). A slave oscillator (normally coupled to a pacemaker) drives a circadian rhythm of dark induction (the arrow indicates the start of the light-induced rhythm) such that if darkness exceeds a certain critical period (for example the increasing night lengths of autumn), diapause will be triggered. In darkness the slave oscillator uncouples from the pacemaker and rapidly damps, leading to the loss of the dark induction rhythm. Under a Nanda–Hamner protocol the periodic arrival of light will induce coupling between the master pacemaker and slave oscillators and thus the reappearance of a rhythm of dark induction. When the night length overlaps with the dark induction rhythm, a diapause response will be triggered. In this example, darkness between 12 and 20 hours after dawn will induce diapause (grey box). If the insect is exposed to light during this period, diapause will be inhibited. Under a Nanda–Hamner protocol of 6 hours light and 6 hours dark (6:6; period length ($T$) = 12 hours), the dark inductive phase is induced by the first 6 hours of light, and the dark inductive phase is positioned approximately 12–20 hours after dawn. But under a photoperiod of LD 6 : 6, light will always fall on the dark inductive phase and there will be no diapause. Under a photoperiod of LD 6 : 30 ($T$ = 36 hours), the dark inductive phase is again induced by the first 6 hours of light and positioned approximately 12–20 hours after dawn. The 30 hours of darkness will overlap with the dark inductive phase and stimulate diapause. The rhythm rapidly damps and disappears under the prolonged darkness, but with the arrival of light the pacemaker and slave re-couple and initiate a rhythm of dark induction. The same phenomenon will occur under any photoperiod with an extended dark period, such as that seen in the last example with a photoperiod of LD 6 : 42. An hourglass-like response, as illustrated in Figure 5.8, could therefore (in theory) be the product of a heavily damped circadian oscillator.

eggs in the ovary indicates diapause, and this response has been used as an assay to study the involvement of clock genes in diapause. The best evidence we have of such involvement comes from very recent collaborative studies by Bambos Kyriacou at the University of Leicester and Rodolfo Costa at the University of Padova.

They have discovered that the circadian clock gene *timeless* comes in two naturally occurring versions, *ls-tim* and *s-tim*. Those flies with the *ls-tim* gene are more likely to show diapause in response to short days and low temperature than *s-tim* females. Although these studies link the *tim* gene to diapause, it remains unclear whether the variant forms of *tim* are modifying the photoperiodic timer itself or perhaps some other non-timing component of diapause such as the photosensitivity of the timer. Despite all the research, the best that can be said at present about the molecular mechanism underlying the timing of insect diapause is that it is still something of a mystery.

Another mystery is the finding that insects also seem to have a circannual timer too! It has been known for more than 40 years that the carpet beetle (*Anthrenus verbasci*) has a circannual free-running period of about 41 weeks that is temperature-compensated and can be entrained by photoperiod. Recent work has confirmed and extended these observations (Nisimura & Numata, 2003), providing unambiguous evidence for the existence of a circannual timer in at least one species of insect. However, as with birds and mammals, we are no closer to understanding the whereabouts and the mechanism of this circannual clock.

# 6

# TIMING MIGRATION

Should I Stay or Should I Go

THE CLASH, FROM THEIR ALBUM *COMBAT ROCK* (1981)

When we think of migration we usually think in terms of the annual comings and goings of huge flocks of birds crossing deserts, oceans or mountains, or Pacific salmon leaping the rapids of an Alaskan river and avoiding the jaws of a grizzly bear. But migration is a term used to describe a broad range of movements ranging from the small-scale passage of micro-organisms through the soil, the large daily vertical movements of plankton in the ocean, and the foraging trips of albatrosses (*Diomedea* spp.) for thousands of kilometres. For our purposes it means behaviours that show a seasonal movement of populations between regions where conditions are alternately favourable and unfavourable, including one region in which breeding occurs. Animals migrate seasonally when they benefit in terms of survival and reproduction.

The seasonal movements of large mammals were observed by our forebears of the Upper Palaeolithic thousands of years ago on the plains and passes of Europe. It was their understanding of the regularity of these events and their ability to anticipate the animals' movements that gave them a competitive edge when they set their ambushes. In Biblical times the prophet Jeremiah talked of 'The stork in the heavens knoweth her appointed

120

time; and the turtledove, and the crane, and the swallow, observe the time of their coming.' The ancient Greeks and Romans also wrote of migration. We have watched animal migration for much of our existence, but despite aeons of observation it remains enigmatic and poorly understood. When Gilbert White wrote his *Natural History and Antiquities of Selborne* in 1789, many were convinced that swallows (*Hirundo rustica*) spent the winter hibernating in holes or under the mud of local ponds. White himself seems to have been undecided. There are folklore stories of fishermen in northern waters pulling in a mixed catch of fish and hibernating birds. The idea that even tiny birds might fly long distances halfway round the world and back again was simply not credible to peoples who lived and died close to where they were born.

Animals fly, swim, walk, crawl and even drift in their frantic need to migrate. The evolutionary origins of migration in different species are hotly debated. In some species much of this effort is to acquire the considerable resources demanded for successful breeding, but in others it may be the avoidance of harsh seasonal conditions and/or predators and parasites. Many make round trips or 'loop migrations' (Figure 6.1), in which individuals return to the breeding area where they were born, passing through non-breeding areas that differ on the outward and return journeys. For others it is a one-way trip. Many insects go from where they were born to the next breeding area, where they produce the next generation and die (Dingle & Drake, 2007). Birds travel prodigious distances. Arctic terns (*Sterna paradisaea*) fly some 32,000 km each year as they travel from their winter sites in south polar regions to their north polar breeding sites and back again. This route back and forth may involve the circumnavigation of the Atlantic or Pacific Oceans depending on the sub-population (Figure 6.1B). The bar-tailed godwit (*Limosa lapponica baueri*) manages a 11,500-kilometre migration non-stop! A tracked godwit flew from Alaska back to New Zealand in eight days.

When populations explode in numbers and outstrip resources, as happens with lemmings and other rodents, they simply leave the area. These are not migrations but irregular dispersals with no return, and represent mass emigration or irruption away from an area. In the same way, the mass movement of the so-called migratory locusts of North Africa

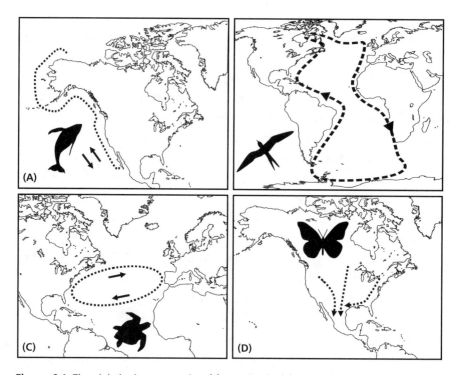

**Figure 6.1** The global migratory paths of four animals. (A) Grey whale (*Eschrichtius robustus*), which follows broadly the same outward and return journey. (B) The loop migration of the Arctic tern (*Sterna paradisaea*), the world's long-distance champion. The birds can travel over 36,000km a year. (C) The loop migration of the loggerhead sea turtle (*Caretta caretta*). Loggerheads undergo one of the longest marine migrations. Hatchlings that emerge on the beach along the east coast of Florida migrate offshore to the North Atlantic gyre, the circular current system that encircles the Sargasso Sea. Young loggerheads remain for several years in the relatively safe open-ocean gyre system before returning to the coastal feeding grounds off the southeastern USA. (D) The one-way migration of the monarch butterfly (*Danaus plexippus*), which can fly 120 km in a day.

(*Locusta migratoria*), which can cause such devastation to crops, are actually irruptions, whereas the North African desert locusts (*Schistocerca gregaria*) make true migrations between their winter and summer breeding grounds.

Much of the flying, swimming, walking, crawling and drifting is, nevertheless, migratory and is associated with finding a suitable place to breed or survive. The new destination often provides a rich and abundant food source to feed the new young and/or a relatively safe place from predation. Green turtles (*Chelonia mydas*) migrate more than 2,000 km across open ocean from their feeding area off the coast of Brazil to their nesting area on the isolated Ascension Island, a tiny island less than 10 km wide. By synchronising their breeding, the near-simultaneous hatch of hundreds of thousands of young overloads or 'gluts' the relatively few predators that can live on Ascension. Many of the newborns get eaten, but far more make it to the sea.

The northern elephant seal (*Mirounga angustirostris*), starts the birthing period around Christmas at the breeding grounds on the isolated beaches along the coast of California. The females remain on the beach for a total of about 40 days, whereas the males stay for up to 100. Neither males nor females feed while they are on land, and both sexes lose about 30 per cent of their body weight during the breeding season. After breeding, males move north and travel to areas along the continental margin, ranging from coastal Oregon to almost 5,000 km away in the Aleutian Islands in western Alaska. Here they stay before heading back to the rookery.

The females feed in the deep waters of the open ocean of the northeastern Pacific. Although the males are vulnerable to attack by great white sharks (*Carcharodon carcharias*) and killer whales (*Orcinus orca*) as they forage along the food-rich continental margin, females are less likely to encounter these predators in the open ocean. Males migrate farther and take more risks than the females, but the reproductive pay-off for those males that survive and breed is huge. One study found that as few as five males out of 180 in the colony were responsible for between 50 and 92 per cent of observed matings with 470 females (Clutton-Brock, 1998).

Northern elephant seals, like many other migrants along the Pacific coast, can be very badly affected by changes in climate. During the 1997–98 El Niño, the mortality of northern elephant seal pups in California was around 80 per cent. Severe storms, elevated sea levels, heavy rains and very high tides combined to submerge colonies and wash away pups. Such events are predicted to increase with climate change (Mathews-Amos & Berntson, 1999).

Perhaps the most dramatic example of migration to an almost predator-free breeding site is the winter trek of the emperor penguin (*Aptenodytes forsteri*) of some 50–200 km over stable pack ice onto the Antarctic ice shelf. In early April the penguins shuffle south while most other Antarctic birds head north and away from the winter. Emperor penguins break all the rules, laying their eggs in autumn and incubating them for four months through the winter. The female lays one 450-gram egg in June. Before leaving on her return trek to the sea to feed on fish, krill and squid, she transfers the egg to the feet of the male, who incubates it in his brood pouch for about 65 days. With no food, he survives on his fat reserves. The temperature in the brood pouch is some 80°C warmer than the midwinter temperature outside (−50°C), and if the egg should touch the ice, the chick inside would die very rapidly.

The breeding colonies are usually in sheltered areas where ice cliffs and icebergs protect against winds that can reach 200 km/h. Emperor penguins do not build nests, and this allows the birds to huddle close together, providing some protection from the cold. When the female returns she finds her mate among the hundreds of fathers by using his call to track him down. She then takes over caring for the chick, feeding it by regurgitating the food that she has stored in her stomach. The male, having lost approximately 45 per cent of his body mass, then leaves for the sea, which is a slightly shorter trip as the melting ice decreases the distance between the breeding site and the open sea. After another few weeks, the male returns and both parents look after the chick. As the weather becomes milder, the chicks huddle in a crèche for warmth and protection, still fed by their parents. It is a ponderous process, but there are few predators in this desolate environment, primarily the southern giant petrel (*Macronectes giganteus*) and the south polar skua (*Stercorarius maccormicki*). Finally the chicks and their parents return to the open sea and feed in the food-rich coastal areas around Antarctica (Barbraud & Weimerskirch, 2001).

Birds are the great migrators. Approximately 40 per cent of the bird species that breed in Britain do not spend the winter in the UK but migrate south, some to southern Europe, others much farther. The swallow (*Hirundo rustica*) may migrate as far as the cape of southern Africa. These are risky journeys and the energy costs are huge (Berthold *et al.*, 2003). Despite this, 65 per cent of all bird species migrate. Even with the high mortality on the journey, migration for the swallows and hundreds of other species is obviously a successful strategy. But quite why this should be is more complex than it might appear. In the higher northern latitudes there is little point in staying over winter as there is little in the way of vegetation and not much time to find it in the short days. But many birds that leave their winter quarters in lower latitudes and travel south could probably remain.

The higher northern latitudes provide an abundance of seeds, fruits and insects in spring and summer. Ian Newton speculates that if no birds migrated to higher latitudes, then any bird that did move there would be able to exploit the rich under-utilised resources and produce more young than if it had remained in lower latitudes and competed with the resident birds. It would have a large selection advantage. He has summarised the position in the following way (Newton, 2008):

> Whereas the advantage of autumn migration can be seen as improved winter survival, dependent on better food supplies in winter quarters, the main advantage of spring migration can be seen as improved breeding success, dependent on better food supplies in summer quarters.

Before they take off, migratory birds run through a repertoire of behavioural and physiological changes. A marked increase in appetite and food consumption, termed hyperphagia, begins about two to three weeks before migration and persists throughout the migratory period. This is accompanied by an increase in the efficiency of fat production and storage. As a result, a migratory bird can increase its body weight through fat deposition by as much as 10 per cent per day, although it is usually 1–3 per cent. The ruby-throated hummingbird (*Archilochus colubris*), which flies across the Gulf of Mexico, and the sedge warbler (*Acrocephalus schoenobaenus*), which migrates from Britain to West Africa, lay down huge fat reserves relative to their size. This is necessary because both these species complete their migration in a single sustained flight. Species that carry smaller loads of fat or fly longer distances typically fly in a series of stages, stopping to feed and put on weight along the route (Åkesson & Hedenström, 2007).

Additionally, the pectoral muscles become larger and well supplied with enzymes necessary for the oxidation ('burning') of fat. Fat is the best fuel because not only is it lighter and less bulky than carbohydrates or protein, but it also supplies twice as much energy per unit of mass compared with protein and glycogen.

One of the most remarkable all-round performances is shown by the blackpoll warbler. Its over-water flight to South America keeps it continuously in the air for 80–90 hours. This has been likened in humans to the metabolic equivalent of running a four-minute mile pace for 80 hours. If a blackpoll warbler were burning petrol instead of reserves of body fat, it would get well over 300,000 km to the litre. A typical blackpoll warbler at the end of its breeding season weighs about 11 grams. In preparing for its Atlantic trek, it may accumulate enough fat reserves to increase its body weight to 21 grams. Given an in-flight fat consumption rate of 0.6 per cent of its body weight per hour, the bird then has enough added fuel for approximately 90 hours of flight for a journey that, under fair conditions, requires about 80–90 hours. The 14 grams of fat in a single Snickers bar would provide one and a half times the amount of energy necessary for the blackpoll warbler's flight from New England to South America (Deinlein, 1997).

Readying for migration often involves a switch in diet. Many autumn migrants switch from insects to a diet of berries and other fruits. In autumn, insect numbers are fast decreasing and fruits (high in carbohydrates and lipids) are becoming abundant. Flight muscles often become more extensive before migration, to provide additional lift to carry these new fat deposits. However, as Theunis Piersma of the University of Groningen has explained, it is more complicated than just eating and storing and burning fat. Whereas penguins, for instance, shrink their guts during a prolonged fast in the breeding season, migratory birds often shrink their livers, stomachs or intestines in anticipation of their journey. Piersma discovered in the mid 1990s that, before migrating, bar-tailed godwits eat enough to make fat account for more than half of their body weight, but they also shrink their viscera. Shrinking their guts and other organs by half or more reduces energy consumption. Their guts then grow after they begin eating again. Piersma has a pithy description of the process: 'Guts don't fly' (Piersma & Gill, 1998).

Most migratory species are solitary for much of the year but start to flock together before, or during, migration. This social behaviour seems to provide improved predator avoidance, food finding and orientation. Some species such as geese, ducks, swans, pelicans or cranes fly in a V-shape, which seems to improve flight aerodynamics and reduces energy expenditure (Nathan & Barbosa, 2008).

Another radical behaviour change is a shift from being active exclusively during the day to flying at night. This shift from diurnal to nocturnal behaviour occurs in many species during migration, including most shorebirds and songbirds. The possible advantages of flying at night include decreased vulnerability to predators, reduced threat of dehydration or overheating, a greater likelihood of encountering favourable winds, and possibly easier navigation. At night the air mass is more stable; during the day, hot air and more variable wind directions tend to dominate. Flying at night also leaves more time during the day to forage.

Caged migratory birds have a seemingly odd behaviour that has provided one of the keys to investigating migration. Since the mid eighteenth century, bird fanciers have known that usually diurnal migratory birds show an overall increase in activity at night and excessive hopping at about the same time of the year as non-captives prepare for migration (Farner, 1950). In the 1930s German ornithologists called this surrogate migratory behaviour 'Zugunruhe'. In 1949 Gustav Kramer used this observable and measurable restless behaviour as a marker for migratory activity when he demonstrated that red-backed shrikes (*Lanius collurio*) and blackcaps (*Sylvia atricapilla*), placed in cages under the night sky, often tended to orient their zugunruhe towards the side of the cage corresponding to the normal migration direction. Microswitches mounted underneath the perches are used as one method of recording the bird movements. Another technique involves keeping a bird in a cone-shaped orientation cage lined with white paper and with an inkpad at the bottom and a wire mesh lid (Figure 6.2). Every time the bird hops forward onto the sloping sides, its inky feet leave a mark.

This laboratory technique provided the means of examining in controlled conditions one of the big questions in bird migration: how do they time their seasonal activities? If birds undertake a repertoire of activities in advance of take-off, they need a way to orchestrate the sequence of changes in diet, fat storage, organ reduction, and so on, and to synchronise the overall timing of these activities to their intended departure. Further, they have to time their departure so that they arrive at the migratory location at the time that maximises their chances of successful breeding in spring and their safe return in autumn.

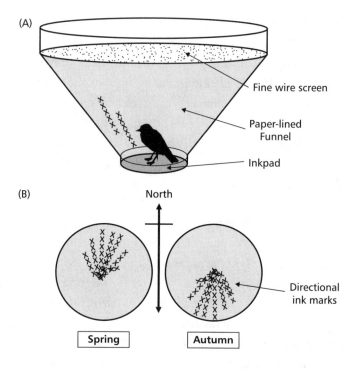

**Figure 6.2** Orientation of migratory restlessness (zugunruhe) in the garden warbler (*Sylvia borin*). (A) The cone-shaped orientation cage has an inkpad at the centre and a wire mesh lid. Every time the bird jumps from the inkpad towards the edge, its feet leave a mark. (B) The direction of marks are oriented in a northerly direction in spring and towards the south in autumn.

In his earliest work, Ebo Gwinner noted an annual, spontaneous development of zugunruhe in captive willow warblers (*Phylloscopus trochilus*) held in constant conditions of light and dark. This provided the first unambiguous evidence for the involvement of an internal circannual clock in a seasonal timing event in birds (Ted Pengelley independently made a similar finding in ground squirrels). Gwinner's work was a confirmation of the remarkable speculation of Count Johann von Pernau, who in 1702 had concluded that birds in higher northern latitudes that begin their migration south relatively early in the year in July and August are not simply driven away by hunger and cold; instead there is a hidden force driving them to depart at the right time (Slater *et al.*, 2003).

The willow warbler spends much of the year in the equatorial regions, migrating north in the spring and then travelling back in the autumn. A complete moult of the feathers occurs in the summer after the northern migration and then again in the winter after the migration south. Rhythms of zugunruhe and moult were found to continue in captive willow warblers maintained on an artificial lighting schedule of 12L:12D for 28 months. The average period for this event was approximately ten months (Gwinner, 1967). Rhythms of zugunruhe and moult have been found to persist for at least ten cycles in many bird species kept under constant conditions so that there is neither contraction nor expansion of the photoperiodic signal. This is convincing evidence that migration is under the direction of a circannual timer (Gwinner, 1986).

Remarkably, the amount of zugunruhe exhibited by birds is both species-specific and population-specific. German blackcaps migrate some 1,000 km, whereas blackcaps from the Canary Islands are non-migratory. When the two populations were studied in the laboratory in the autumn, the German blackcaps showed high levels of nocturnal zugunruhe for up to 150 days, whereas the birds from the Canary Islands showed a small amount of increased nocturnal activity for a period of a little over 40 days. When a hybrid bird was created by crossing birds from the two populations, the offspring showed an intermediate level of zugunruhe, suggesting that the migratory programme of blackcaps, and perhaps most migratory birds, has a strong genetic component that encodes how long birds migrate for, and thereby how far they travel (Berthold & Querner, 1981).

The direction of migration is also embedded within a seasonal temporal programme in birds. Working with the garden warbler (*Sylvia borin*), Gwinner hand-raised birds and then maintained them under a constant 12L:12D photoperiod and constant temperature for their first autumn and spring migratory seasons. At regular intervals, the birds were tested in circular orientation cages for the directional preferences of their nocturnal zugunruhe. During these tests, the birds were exposed to the local magnetic field of the earth but had no view of the sky. The birds showed a highly significant preference for a southerly direction in autumn and for a northerly direction in spring. Gwinner concluded, 'These results support the hypothesis that the reversal of the migratory direction results from spontaneous changes in the preferred direction relative to external orienting cues that are controlled by an endogenous circannual clock' (Gwinner & Wiltschko, 1980). Similar results have been found in a range of migratory species including blackcaps (*Sylvia atricapilla*) (Helbig *et al.*, 1989) and pied flycatchers (*Ficedula hypoleuca*) (Gwinner, 1996b).

This temporal circannual programme is remarkably stable. Captive birds can be exposed to conditions of complete darkness, high humidity, simulated rain, and even mild food restriction, but the timing, amount and direction of zugunruhe go on normally. It is only when birds are subjected to cycles of near starvation followed by re-feeding that the amount of zugunruhe begins to drop. The amount of zugunruhe can also be increased under sustained food restriction. But the only environmental factor capable of altering the timing of the migratory programme is photoperiod. Exposing birds to short photoperiods advances the time of the autumn migration, and long photoperiods advance spring migration. As a result, photoperiod seems to be the primary entraining agent for the circannual rhythms of migration (Gwinner, 1996a). The daily timing, as opposed to the seasonal timing, of zugunruhe is under circadian control. Blackcaps (*Sylvia atricapilla*) showing appropriately timed seasonal nocturnal zugunruhe under a spring or autumn light:dark cycle were transferred to constant conditions of light and temperature. A pattern of activity and rest persisted but drifted (free-running) with a circadian period of about 25.5 hours (Gwinner, 1996a).

Circannual clocks provide the major influence in the timing of migrations in both directions. In equatorial regions, the photoperiodic signal is weak and a poor predictor of seasons, so it makes sense that birds use a circannual timer to regulate their migration from equatorial to higher latitudes. But birds living in the tropics have not abandoned photoperiodic timing entirely. The spotted antbird (*Hylophylax naevioides*) lives in the lowland moist forest of central Panama (98° N 79° W). In laboratory conditions, birds were able to respond physiologically and behaviourally to an increase in photoperiod of as little as 17 minutes (Hau *et al.*, 1998).

It is perhaps more surprising that birds use a circannual timer to time migrations from high to low latitudes, but they do use it together with photoperiod. In the higher latitudes the photoperiodic signal is strong and could, as in the case of reproduction, time the migration south. Yet the circannual clock again seems to be the dominant timer. One explanation is that the circannual clock prevents 'photoperiodic confusion'. As soon as the bird moves to lower latitudes, daylength will change. A circannual reference system could override such mixed signals.

Not all species of birds respond in the same way to seasonal change. Some are rigid in their timing, such as the 'miracle' of the swallows of San Juan Capistrano, which leave the small Californian mission on 23 October, the day of San Juan, each year to winter in Argentina, from whence they return en masse on 19 March. Although legend has it that the birds have only been one day late in more than 200 years, it is perhaps better to accept that they migrate to a tight, but not exact, schedule. But in most cases there is considerable variation in the sequence and duration of migration, breeding and moult, not only between species but even in different populations of the same species. The interaction of an endogenous circannual rhythm, the photoperiod and daily circadian cues provides animals and migratory birds in particular with what Gwinner described as the major basis for their 'orientation in time' supplementing their 'orientation in space' (Gwinner, 1996a).

Although the flexible combination of endogenous timer synchronised by photoperiod to seasonal cycles and sensitive to local environmental cues works, nobody has yet found the elusive circannual pacemaker and nobody knows how it works. Our ignorance about birds is as profound as it is about mammals. However, there are suggestions that the pattern of melatonin from the pineal may be involved. The amplitude of melatonin release at night is lower during nocturnal zugunruhe (Gwinner, 1996a, 2003). It is unclear whether this is cause or effect – is the melatonin level lower because activity is higher, or does the decrease in nocturnal melatonin level trigger zugunruhe?

Although several models for generating a circannual rhythm have been suggested, we are really no further forward today in understanding the mechanism than we were more than 20 years ago when Gwinner wrote, 'The problems stem partly from our almost complete ignorance of the physiological processes involved in generating circannual rhythmicity' (Gwinner, 1986).

Migration is a separate activity from reproduction, and in many regards they are energetically opposed. Before the bird departs the breeding ground and flies south, its reproductive system has already regressed under the influence of photoperiodic mechanisms. The circannual clock then times migration, while movement to the breeding ground requires that birds be reproductively competent at or shortly after arrival. This means that the reproductive programme needs to be activated during the journey. This makes for a tricky timing issue to get the balance right between migration and reproduction. The bird cannot be burdened with an increasingly heavy set of reproductive organs while it is still flying long distances. Yet it needs to have the appropriate reproductive behaviours, which include territoriality and courtship, in place virtually the moment it lands at the breeding ground. The birds obviously get this right. Testosterone is involved in kicking off reproduction, but precisely how this timing is achieved remains unknown.

For most species, pairs that start to breed early in the spring fledge a greater number of young. Early arrivals generally secure better territories in terms of the abundance of food that they contain, and an early start often means that a multi-brood species pair has the time to produce and rear a second clutch in the same season.

It could be that the earliest birds to arrive at the breeding sites are in better condition than later birds, some of which arrive up to a month later. But other factors may be important. There is good evidence that the quality of winter habitats may be an important determinant of reproductive success for migratory birds. In a study of American redstarts (*Setophaga ruticilla*), the early-arriving, good-condition birds had wintered in better tropical habitats than the birds that arrived later (Norris *et al.*, 2004). So it is important to consider all parts of the life cycle when considering breeding success.

The timing of zugunruhe is one part of the problem of migration; another critical issue is to know in which direction to go. Birds are aware of the landscape over which they fly, and they seem to use landmarks for orientation purposes. Radar images of migrating birds exposed to strong crosswinds show the flock drifting off course, but not if the flock migrates parallel to a major river. It seems that these birds can use the river as a reference to shift their orientation and correct for drift so as to maintain the proper track. In the same way, migrating hawks trying to locate updraughts along the north shore of Lake Superior or the ridges of the Appalachians must be aware of the terrain below if they are to take advantage of the energetic savings afforded by these topographic structures.

Physical landmarks in the environment are clearly very useful as a means of navigation, but only if they are visible and the bird has been there before. For cranes, swans and geese that migrate in mixed-age groups, the young could learn the geographic map for their migratory journey from their parents or other conspecifics. But most birds do not migrate in family flocks, and on their initial flight south to the wintering range or back north in the spring they must use other cues. Several 'environmental compasses' are used for orientation, including the position of the sun and the stars, and even the earth's magnetic field.

If the sun is to be used as a navigation aid, there must be a way of compensating for its apparent movement across the sky. If a bird in the northern hemisphere wishes to fly south, it needs to keep the easterly sun on its right in the morning at 6.00 a.m., fly towards the southerly sun at noon and keep the westerly sun to its left in the evening (Figure 6.3).

In the 1950s, Gustav Kramer was the first to show that the sun was used by different bird species to navigate and that they could compensate for the sun's apparent movement across the sky. Kramer's initial experiments were on European starlings (*Sturnus vulgaris*), which he trained outdoors to fly off in a particular compass direction to get a food reward. He suspected that the birds were using the sun's position in the sky (azimuth) as a compass. To test his hypothesis he moved the starlings indoors, replacing the sun with an electric light as a direction giver. As before, the birds inside would similarly move off in a particular compass direction for the food reward, but this time using the light-bulb as an orientation cue. The electric light was fixed in position; remarkably, although the birds would move off in the direction of the food, they added 15° of arc (anticlockwise) each hour to the angle they made to the artificial sun.

Kramer concluded that the birds must 'know' that the angular velocity of the sun's azimuth is approximately 15° per hour, that they have some clock to compensate for the moving sun compass, and that they 'knew' that they were in the northern hemisphere (Kramer, 1952). Shortly after this pioneering work – which was one of the earliest demonstrations of a circadian clock

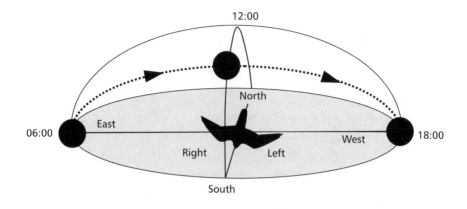

| Environmental Time | Circadian Time | Angle Flown |
|---|---|---|
| 06:00 | 06:00 | 90° Right |
| 12:00 (noon) | 12:00 | 0° Front |
| 18:00 | 18:00 | 90° Left |

**Figure 6.3** Time-compensated sun-compass orientation enables animals, in this case a bird, to compensate for the sun's apparent movement (black spheres) across the sky. An internal biological (circadian) clock provides the time information for the bird to adjust its direction by about 15° per hour to compensate for the moving sun. In this example the bird has been outside and so real 'clock' time and circadian time are synchronised. To fly in a southerly direction, this bird in the northern hemisphere must keep the rising sun in the east on its right in the morning at 6.00 a.m., fly towards the southerly sun at noon and keep the westerly sun to its left in the evening.

– Kramer tragically died in 1959, when he fell while searching for clutches of rock pigeon eggs for his hand-rearing experiments. In follow-up experiments in the late 1950s Klaus Hoffman, a colleague of Kramer's, shifted the circadian rhythm of birds with respect to the sun by exposing them to an artificial L:D cycle. He found that the orientation of the birds shifted approximately 15° of arc for every hour that the birds' clock had been shifted. This deviation corresponds precisely to the difference between the actual angular position of the sun and the position predicted by the birds' own internal clock (Hoffman, 1954) (Figure 6.4). Many assumed that sun-compass orientation is an innate capability of birds. However, in 1981 researchers demonstrated that young birds must experience the sun's arc across the sky

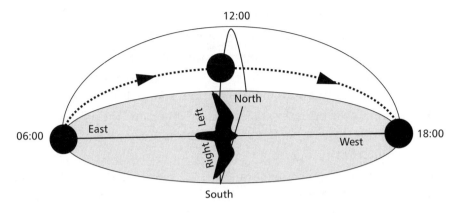

| Environmental Time | Circadian Time (delayed by 6h) | Angle Flown |
|:---:|:---:|:---:|
| 06:00 | 00:00 | – |
| 12:00 (noon) | 06:00 | 90° Right |
| 18:00 | 12:00 | 0° Front |

**Figure 6.4** Time-compensated sun compass orientation depends upon a suitably adjusted circadian clock. Klaus Hoffman placed Birds in an artificial L:D cycle that was delayed by 6h compared to environmental time. Birds were kept in these conditions for a week and their circadian rhythms shifted to the new time. The migratory bird shown (like that depicted in Figure 6.3) would normally want to migrate south but it has had its body clock delayed by 6h. As a result when it is released into the wild at 12.00 noon environmental time, the bird's body clock 'thinks' it is dawn and as a result will keep the sun 90° to its right and instead of flying south it will fly west. At an environmental time of 18.00 the bird's body clock will 'think' it is 12:00 noon. At this circadian time the bird normally flies towards the sun and so the bird continues to fly west. This type of phase-shift experiment has demonstrated that birds, and other animals including monarch butterflies, use their circadian timing system to adjust for the apparent movement of the sun in the sky. This enables them to use the sun as a stable reference to define compass directions (compare with Figure 6.3).

if they are to use it as a compass cue as adults (Wiltschko & Wiltschko, 1981).

The sun not only moves across the sky as the earth rotates, but its apparent position also varies with latitude and time of year. It is not known how animals compensate for this additional variance. Furthermore, many species such as insects, fish and amphibians do not even have to view the sun directly but can infer its position because of the pattern of polarised light in the sky. This is possible because the properties of skylight polarisation change predictably with the changing position of the sun.

Birds also use the stars, when visible, to provide compass information. In 1957 Franz Sauer put birds in a planetarium and projected stars onto the ceiling to mimic the open sky at night. He manipulated the position of the stars by rotating them through 180° such that the orientation of Polaris and all stars around Polaris were flipped into the southern sky. Remarkably, this rotated the direction of the zugunruhe by 180°. He concluded that birds use some kind of view of the stars to orient their zugunruhe. Stephen Emlen provided the proof for this when he published his planetarium experiments on indigo buntings (*Passerina cyanea*) showing that they can orient in a non-directional magnetic field by using directional cues from the stars. Birds learn the star map in their youth and use this information to orient their zugunruhe for their first migration (Emlen, 1970, 1975).

Birds also orient themselves by detecting magnetic fields. In the mid twentieth century the German scientist Fritz Merkel placed European robins (*Erithacus rubecula*) in a large cement cage that screened out all environmental clues; the birds still were able to orient their zugunruhe properly. But when he put the birds into a large steel cage, which interrupted magnetic lines of force around the cage, the birds oriented randomly (Merkel & Wiltschko, 1965). Species as diverse as loggerhead turtle hatchlings and migratory birds can detect and orient within a magnetic field (Lohmann *et al.*, 2007).

Perhaps even more remarkable than animals sensing a magnetic force is the fact that we have some understanding of how they do it. After all, magnetism is not at the forefront of human senses. There seem to be several different mechanisms for magnetoreception in the animal world. One mechanism is based on an iron oxide called magnetite and involves tiny crystals of permanently magnetic material. In pigeons, magnetite crystals have been found in specific locations in the skin around the upper beak (Fleissner *et al.*, 2007). Electrophysiological recordings from the nerves serving these regions show that their activity changes when the magnetic field around the snout is altered, and anaesthetising these nerves interferes with the orientation of migration.

Although migrating herds of large mammals use smell to help them navigate, birds have long been thought to have a poorly developed sense of smell and certainly not one capable of providing navigation cues. However, some species can discriminate olfactory cues quite well and may develop a navigational 'smell map'. If the olfactory nerves of homing pigeons are cut, the birds do not return to their home loft as well as birds whose olfactory nerves are left intact (Bingman *et al.*, 2003).

Birds integrate these multiple cues for accurate navigation. For a migrating bird this overlap is critical, because not all sources of orientation information are equally available at a given time, nor are all sources of information equally useful in a given situation. The different sources of navigational information reinforce each other rather than acting as alternatives. Just as they are flexible in the use of a suite of methods to time their seasonal events, so birds are flexible in their use of a suite of navigational aids. They have to do so, because the conditions they encounter are so variable. When the sun or stars are obscured by clouds, navigating by sensing the earth's magnetic field is very handy. Multiple sources of compass information are clearly adaptive (Lohmann *et al.*, 2007).

Much of the discussion in this chapter has focused on birds because there is a laboratory technique for studying their migration. It is hard to study a herd of migrating wildebeest in the laboratory. But birds are not alone in their ability to navigate long distances. Some populations of caribou or reindeer (*Rangifer tarandus*) make mass migrations of more than 1,000 km, covering up to an astonishing 150 km a day, in search of good grazing. Herbivorous mammals often follow well-established trails using their sense of smell. Studies of salmon indicate that they depend heavily on their olfactory sense to locate and return to their stream of origin (Dittman & Quinn, 1996). Bats, whales and seals use echolocation to navigate in the dark or underwater; in addition, some whales seem to take visual bearings on objects on the shore in their migrations.

But the most spectacular migration of all may be that of the monarch butterfly (*Danaus plexippus*), which travels from North America to Mexico. During the autumn, the North American landscape is dominated by the colour orange, not just because of the spectacular change in foliage but because of the migration of millions of monarch butterflies on their journey south. Some will travel 4,000 km.

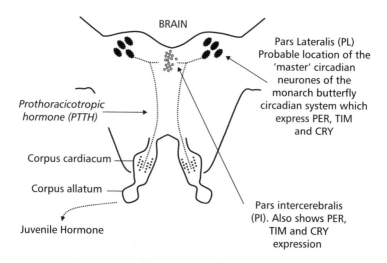

**Figure 6.5** Diagram illustrating the monarch butterfly brain, showing the location of the likely site of the 'master' circadian clock in the brain within the pars lateralis. The pars intercerebralis produces insulin-like neurohormones, which help regulate the release of juvenile hormone. CRY-containing neurones connect the circadian clock in the pars lateralis to the pars intercerebralis.

While migrating, the butterflies are in a state of reproductive diapause, suspending their reproductive effort to conserve all their energy for the trek south (Brower, 1996). These butterflies have built up fat stores before developing an overwhelming urge to fly south. They stop en route to top up their fuel supplies by feeding on nectar. On arrival at a small number of sites in Mexico, reproductive diapause persists until the early spring, when the increasing daylengths break this reproductive block. The butterflies mate and then fly north to the southern USA to lay their fertilised eggs on the spring growth of milkweed plants (genus *Asclepias*). Caterpillars sustained by milkweed give rise to reproductive adult butterflies, which move north, slowly tracking fresh milkweed growth.

Monarchs produce two or three generations of short-lived summer butterflies, the last generation reaching the northernmost reach of their range. Then the decreasing daylengths of autumn trigger the migratory generation to travel south. Steven Reppert, in the Department of Neurobiology at the University of Massachusetts Medical School, has spent much of his long and distinguished career elucidating the mechanisms of the circadian clock of mammals. But in recent years he has switched the focus of much of his work to understand monarch butterfly migration. He argues that this animal is a possible model for understanding the cellular and molecular mechanisms underlying not only the regulation of migration timing mechanisms but also time-compensated sun-compass navigation (Reppert, 2006).

Just eight cells in the monarch butterfly brain near the pars lateralis seem to be the 'master' circadian clock (Figure 6.5). One of the clock proteins, PER, has robust 24-hour oscillations in its production and degradation; it is co-localised with TIM and CRY. A second region of the brain – the pars intercerebralis – also expresses PER, TIM and CRY proteins but these proteins are only weakly rhythmic (Reppert, 2006).

The circadian clock in the paired pars lateralis is almost certainly involved in measuring the shortened daylength and triggering the autumn migration south. In the autumn, juvenile hormone levels decline and butterflies show reproductive diapause. In addition, the decline in juvenile hormone is associated with an increase in longevity from a few weeks in the summer forms to several months in the migratory butterflies. The pars intercerebralis produces insulin-like neurohormones, which seem to help regulate the release of juvenile hormone. CRY-containing neurones connect the circadian clock in the pars lateralis to the pars intercerebralis, and this arrangement may allow the pars intercerebralis clock cells to 'instruct' the pars intercerebralis, and other target neurones in the brain, about the photoperiod and hence the appropriate reproductive response (Reppert, 2006).

The navigational abilities of the monarch butterfly are programmed and not learned. Central to this ability is a time-compensated sun-compass mechanism provided by the circadian clock, which itself is entrained to local time by dawn and dusk. The butterflies continually correct their flight direction relative to the changing position of the sun. Shifting the circadian clock of butterflies by advancing or delaying the light/dark cycle by six hours causes a predictable change in the direction of flight (Froy et al., 2003). In addition, if butterflies are kept in constant light, the molecular clock breaks down and the butterflies' circadian rhythms are lost. In these conditions, time-compensated sun-compass orientation is destroyed.

The butterflies use the pattern of polarised light in the sky rather than the sun itself as their compass cue, and this polarised light seems to be detected by ultraviolet photoreceptors arranged around the dorsal rim of the eye (Sauman *et al.*, 2005). It is also possible that monarch butterflies have magnetoreceptors that use the flavoprotein cryptochrome (CRY) as the basis of a magnetosensitive mechanism (Froy *et al.*, 2003).

Migration is one of the most striking adaptations in the animal world to seasonal change. How animals time their migration, obtain the energy for their journey and navigate its course, constitute one of the great feats of nature. Our understanding of the molecular, cellular and even physiological mechanisms that drive this process, despite more than 50 years of investigation, is bereft of detail. This paucity of information reflects the complexity of the migratory process, the diversity of vertebrate and invertebrate taxa, the diverse modes of movement over land, through water and in the air, and finally the fact that few but the most dedicated scientists are willing to spend year upon year trying to tease these processes apart.

Our understanding of the seasonal timing mechanisms in animals is poor. Humans, too, are seasonal – often in surprising ways – but if we little understand animals, our knowledge of how the response to seasonal change is generated in humans is rudimentary.

# 7

# SEASONS AND HUMAN EVOLUTION

Live in each season as it passes; breathe the air, drink the drink, taste the fruit, and resign yourself to the influences of each.

HENRY DAVID THOREAU (THOREAU, 1906)

The regular seasonal change that has been so important in the evolutionary history of plants and animals has also affected us. Even in the equatorial rainforests where our odyssey began, seasonal effects played their part in shaping the environment in which our early ancestors struggled to survive and reproduce. As our forebears moved into ever higher latitudes, they encountered more extreme changes in weather as autumn turned into winter and spring into summer months. The seasons may not have made us, but they helped to shape us.

Our evolutionary journey began in Africa around eight million years ago when our earliest ancestors split away from their close cousins, who would become the chimpanzees and other great apes, and headed off in what was to become our direction. Early to mid-Miocene Africa had a wetter climate than today. Uninterrupted tropical rainforests spread across equatorial lowlands from the Atlantic to the Indian oceans. But about nine million years ago, tectonic forces had begun the formation of the East African rift valley (Cane & Molnar, 2001). Mountains formed on the west side of the rift and created a rain shadow over East Africa by impeding the easterly flow of rain clouds. Alongside this local effect, changes in the pattern of ocean currents led to a simultaneous global cooling. The overall result was that East Africa became significantly drier than it had been.

The consensus until not too long ago was that the common ancestors of modern chimpanzees and humans were split into two geographically separate populations by these geological events. The primate populations of the tropical rainforests of West Africa evolved into modern chimpanzee species. On the increasingly open and dry habitats of eastern and north central Africa, evolutionary adaptation of the primates to a changing climate, followed by adaptive radiation, led to the eventual emergence of modern humans.

However, in 2002 the discovery of fragments of *Sahelanthropus tchadensis* (better known as Toumaï), by Michel Brunet of the University of Poitiers and his team in northern Chad, suggested not only that very early hominins lived in other parts of Africa but also that they were in central Africa a good million or two years before later versions turned up in the east (Vignaud *et al.*, 2002). (Note: hominins is a term used to describe the 'human' subset of the hominids; hominids includes both the human-like and great apes.) That could mean that our early ancestors came out of the forests in what is now Chad and its environs and developed from there, or that a host of still to be discovered ancestral forms emerged at different points in Africa, all except one of whose lineages became extinct as small twigs on the evolutionary bush of which *Homo sapiens* is but an extremity. The fossil record is still far too incomplete to trace the evolutionary paths of our early ancestors.

Toumaï lived six to seven million years ago in what is now the Djurab Desert. This flat, arid region on the southern side of the Sahara was a very different place then. Lake Chad, which nowadays covers some 5,000 square kilometres, could have extended up to 300,000 square kilometres – getting on for the size of France. On the basis of the vertebrate fossils found there, the area around the excavation site in the Toros-Menalla region was one of wooded savannah, fresh water and gallery forests growing along streams and rivers. It was a mosaic of habitats, including desert, near the boundary of a lake. If Toumaï was a very early hominin, and not everyone is convinced that the badly distorted skull is not in reality an early ape, then it was almost certainly a social creature and mainly herbivorous, as were its forest-dwelling primate relatives, with possibly the odd opportunistic scraps of meat coming its way.

*Sahelanthropus tchadensis* was living in the variegated environment near the lake when there was a general trend in Africa to a more marked gradation between a cooler, wet season and a hotter, dry season. It may even have been a more extreme version of the monsoon climate in modern India.

In the tropical rainforest, now as then, much of the food comes from trees. They disperse their seeds by offering fleshy fruits for bats, birds, monkeys and apes to pick direct from the tree and carry away, to be deposited at some distance pre-packaged with fertiliser. Fallen overripe fruit and fruit dropped by monkeys are eaten by small forest antelope, pigs and forest elephants. As there are no clear seasons, the fruiting trees such as figs ripen sporadically throughout the year. When a fig tree finishes fruiting, fruit-eating animals must search out the next tree that is coming into fruit. Around 90 per cent of the trees of the mature tropical rainforest use fruit to disperse their seed, so fruit is a major food resource in a tropical rainforest – especially for those able to reach it high up in the canopy (Galdikas, 1988).

Early chimpanzees living in these rainforests looked much like their modern descendants. The ancestral forms lived as chimpanzees do now, mainly on ripe fruits whose seeds they dispersed. But not all tropical forests are equally productive all year round, and even the evergreen wet tropical forest grades into the seasonal tropical forest. Some have pronounced dry periods, and the trees display a kind of deciduous behaviour. This is not quite as constant and reliable an environment. Most chimpanzees can cope with this seasonality because they can find food supplies all year round. They can eat leaves and shoots as well as fruits; they can hunt other primates and are capable of considerable dietary diversity. But living on a high-carbohydrate diet is hard work. Jane Goodall has estimated that during a typical chimpanzee day, 47 per cent of its waking hours are spent eating, while 13 per cent are spent travelling between food sources (Goodall, 1986). Although feeding occurs all day, consumption peaks occur around daybreak until about 9.00 a.m. and also from late afternoon to sundown (Newton-Fisher, 1999).

The more seasonal the forest, the harder it gets and the competitive pressure both within and between species increases. In the dry season, a decrease in resource variety and abundance causes many animals to exploit more of a single food item, or perhaps a greater variety of foods that they may not have sought out before. These might include underground storage organs in plants, nuts, or other specialty food items.

Whereas the ancestors of chimpanzees and apes remained in the seasonal forests, it is possible that other anthropoids, including Toumaï's lineage, would have been forced out by competition into the seasonal woodlands of central and southern Africa, seasonal Ethiopian uplands, and the open woodlands of the deep sandy soils of what is now the Kalahari. As the increasing seasonal intensity put pressure on resources at different times of the year, the anthropoids had to adapt to the conditions to survive. Robert Foley, Professor of Human Evolution at the University of Cambridge, has been at pains to ensure that the vital conclusion to draw when looking at evolutionary patterns is that 'local conditions are of immense importance in determining how animals, including hominins, would respond to seasonal conditions. There was no single hominin strategy' (Foley, 1993).

Toumaï was not a prototype human and may or may not have been a dead end in terms of the evolution of the *Homo* genus. She was basically an ape, but with some novel characteristics. It would be simpler to understand both Toumaï's and our own evolution if we accepted Frans de Waal's point that we are not so much descended from apes as we *are* apes. Linnaeus, so the story goes, assigned humanity its own genus, *Homo*, for no other reason than a desire to stay out of trouble with the Vatican. He knew that the case for a separate classification was weak (de Waal, 2005).

It was a long evolutionary trek from Toumaï or some similar early forms through to recognisably modern humans. One seminal event occurred some time around 1.8 million years ago, when *Homo erectus*, the first of our ancestors to hunt systematically, migrated out of Africa. Tools and remains of this species have been found widely distributed in Europe and Asia. Quite why the migration occurred, after six million or so years of evolutionary development solely in Africa, is still unknown. It may be that there was a general migration of animals, including carnivores, around this time and the *Homo* species went with the flow, so to speak. These early humans were probably nomadic, moving in response to the seasonal transformation of vegetation and the movements of game. Whatever the reason, within a few hundred thousand years, members of the *Homo* genus had reached various sites up to 40° N, at Dmanisi (Republic of Georgia) and Longgupo Cave (central China).

As the early humans expanded their geographical range away from equatorial regions, the availability of resources and the need for shelter were tempered by the seasons. In Steven Mithen's view (Mithen, 2005):

> While these new environments varied greatly from one another, they were all sub-stantially more seasonal than the low latitudes of Africa. If the earliest Homo had mastered low-latitude savannah environments, Early Humans had the capacity to learn about a much wider range of new environments, most notably those of the high latitudes with their very different landscapes, resources and climates.

Seasonal change and its accompanying selection pressure may well have been a key factor in the development of the *Homo* genus.

At some stage, Early Humans learned the controlled use of fire that enabled them to live at the increasingly colder higher latitudes and altitudes. The earlier its use, the easier it is to understand the timing of migrations to higher latitudes; the later its development, the more scepticism there is about the dates of entry into non-equatorial zones. A time 300,000 to 500,000 years ago was the best guess, but recent work in Israel suggests that fire was being used some 800,000 years ago (Goren-Inbar *et al.*, 2004). This find has also given more credence to evidence for the controlled use of fire by *Homo erectus* proposed for burnt bones at the site of Swartkrans in South Africa dated to 1.5 million years ago (Brain & Sillen, 1988) and for patches of baked dirt at Chesowanja in Kenya at 1.4 million years (Gowlett *et al.*, 1981).

As well as providing protection against wild animals, fire would have enabled hominins to cook their food, stay warm during the winter and possibly improve their weapons. Fire made it possible for hominins to use caves as more than short-term resting areas during the day. Caves are usually cold and damp, making them uncomfortable and dangerous for permanent use. With fire, not only could hominins see inside the cave but they could build up large fires in cave entrances to keep predators at bay and themselves warm

(Nicolas, 2004). They could also cook and, as in Plato's famous allegory, the shadows on the wall can give rise to contemplative thoughts about the mysteries of human existence.

Although the controlled use of fire postdated the very first migrations out of Africa, its advent enabled early hominins to colonise vast new tracts. The higher latitudes of Asia were occupied perhaps a million years before those in Europe, possibly because, again in Steve Mithen's words, 'the degree of seasonal variation in the Pleistocene in higher European latitudes was still beyond the cognitive capacities of the Earliest Humans to cope with' (Mithen, 2005). The time scales may have varied, but the ubiquitous distribution of humans around the globe is largely a result of an understanding of the timing of seasonal variation and an ability to manipulate the environment to withstand and exploit the consequences of predictable, rhythmic change.

*Homo erectus* became extinct about 200,000 years ago. Early forms of *Homo sapiens* – the Early Modern Humans – appeared around the same time, and by the Sangamon interglacial (100,000 years ago) existed in two varieties, *Homo sapiens sapiens* (Cro-Magnon, or Modern Human) and *Homo sapiens neanderthalensis* (Neanderthals).

Outside Africa and the Middle East, there are no clearly dated fossils of *Homo sapiens sapiens* older than about 50,000 years, suggesting that the last human migration out of Africa was very recent. About 50,000–60,000 years ago a group of perhaps a thousand or so of these Modern Humans probably crossed out of Africa at Bab el Mandeb, the outlet of the Red Sea to the Indian Ocean. Over the years small bands split off and started the journeys to the far corners of the globe. These fully modern people soon developed trading patterns in 'art' objects. Shells from the Mediterranean have been found at excavations at Perigord in France. Man-made adornments such as beads and necklaces, bracelets and pendants were produced at 'factories' at Abri Blanchard, Abri Castanet and La Souquette. David Lewis-Williams, the South African authority on Palaeolithic painting, notes that trade implies social complexity and communication (Lewis-Williams, 2002):

social integration beyond a nuclear family comprising parents and children is also evident in the organisation of hunting, especially of migrating herds of bison, horse and reindeer. The steep-sided valleys of the Vézère and the Dordogne, for instance, channelled herds of animals migrating from the Massif Central down to the plains that lie to the west. To take advantage of animal migration, people had to predict the time and place best suited to hunting and then to organise parties to be present at the right times and to perform different but complementary functions. Upper Palaeolithic people were also able to predict the early spring salmon runs when these fish swim upstream to spawn. Being at the right place at the right time meant a great harvest of fish that they could dry and store.

Modern humans quickly replaced the Neanderthals and settled in what was an increasingly harsh landscape. The ice sheets were moving south and would eventually reach a latitude running from southern England to northern Germany. Although it was cold, huge herds of large grazing mammals migrated across the treeless landscape. These early Europeans have been described by Ian Tattersall, of the American Museum of Natural History, as

> hunters and gatherers: people who lived off the resources available on the landscape. ... For skilled hunters with all the cognitive powers of modern humans, the abundant fauna of the open steppic landscape was an incomparable resource to be exploited, sometimes with relatively little effort.

Tattersall has surveyed the evidence and he is certain (Tattersall, 2004) that

> the Cro-Magnons carefully monitored their prey over the seasons of the year: animal depictions sometimes show bison in summer moulting pelage, stags baying in the autumn rut, woolly rhinoceroses displaying the skin fold that was visible only in summer, or salmon with the curious spur on the lower jaw that males develop in the spawning season.

As *Homo sapiens sapiens* travelled and navigated their way around, whether for hunting, fishing or trade, they needed to be aware of seasonal patterns. The evidence from finds of large piles of animal bones suggests that these Modern Humans were much better at predicting the movements of animals than Early Humans. Because they understood and could better predict seasonal change from the observation of animal and plant behaviour, they were able to shift from hunting individual and small groups of animals to slaughtering mass herds of reindeer and red deer, which they attacked at critical points on the migration route.

From our beginnings in our African homelands, the availability of resources, the need for shelter and the need for clothing to varying degrees have all been modulated by the seasons. While the focus of our ancestors was on the plants and animals in the immediate environment around them, it seems certain that the small nomadic societies, migrating in response to the seasonal changes in vegetation and the movements of animals, would have noticed the rhythmic changes of the sky. At some point they took advantage of them. They watched for the seasonal appearances of key stars, helping them anticipate the changing seasons. And they would almost certainly have begun using the phases of the moon as a form of calendar. Anthony Aveni has pointed out, 'they knew when to hunt, when to gather and they could certainly tell when the extended light of the moon would come, simply by spotting the first lunar crescent in the west after sunset' (Aveni, 2000). They were also aware, as humans had been for millennia, well before the Palaeolithic transition, that at certain times of the year animals and plants are less abundant than at others. At some point early humans must have begun some form of ritual aimed at maintaining an adequate supply of food. As their societies developed, this ritual behaviour would have been codified and probably timed to phases of the moon because this was and is the most obviously rhythmic body in the heavens.

Lunar cycles have dominated human culture for thousands of years and still do. In 2001, the first ever cricket test match between Sri Lanka and Zimbabwe had to be postponed by one day because of a new Sri Lankan government rule that banned the playing of sport on a full moon. But despite the persistent belief that our mental health and a multitude of other behaviours can be modulated by the phase of the moon, there is no solid evidence that the moon can influence our biology (Foster & Roenneberg, 2008).

Getting the prediction and the timing of seasonal change right would have provided a significant advantage when we were hunter-gatherers. As we started to farm it became crucial. And the early seafarers would have had a spectacular view of the heavens through an unpolluted atmosphere. The experience would presumably have quickly turned into a means of navigation. After all, birds can navigate by the stars.

Our earliest ancestors came out of the equatorial forests and left a mildly seasonal environment for one in which they had to battle continually with the cyclic changes between hot and cold, wet and dry, calm and stormy. Those changes were engines of natural selection. In what has to be informed conjecture, somewhere along the human evolutionary journey across the millennia and across the globe, seasonal change may have led to one of our basic cultural constructs, namely our perception of time. Animals can measure time intervals such as when birds forage for food, but the reflexive conception of a time past, present and future is somewhat different. Thomas Zimmermann of Bilkent University considers (Zimmermann, 2003) that this deeply human concept may have developed because

> a nomadic lifestyle defined, through seasonal travelling, the basic constants of our existence – the change of seasons, dawn and sunset, changing climate, to be at the mercy of heavy rainfall or burning sun – and this affected prehistoric man in a much more intense way than we can imagine today. These constants forced man to follow a rhythm that can be structured into phases of rest and phases of activity.

The higher primates, of which we are one, are diurnal creatures. Our internal biological clocks generate a circadian rhythm synchronised to the daily rotation of the earth on its orbit around the sun. We mask this rhythm in our workaday lives, but 25,000 years ago our Cro-Magnon forebears would have risen and gone to bed, metaphorically speaking, with the sun.

Even though their visual systems would have switched to rod-dominated low-light monochrome vision with the impending onset of dusk, they were not night-time creatures. Our Upper Palaeolithic ancestors lived in a dangerous environment in which being half asleep was a recipe for disaster. Consequently, they were up and alert in the day, and asleep in a shelter at night. They had to know the time of day and likewise the time of year. They had to know when the seasons were changing and what that would mean for their survival and reproduction. To what extent this came from innate circannual rhythms and to what extent it was 'learnt' behaviour that came from understanding the signs that denoted forthcoming seasonal change is arguable, but some of our 'smarter' ancestors exploited the knowledge to obtain food or seek shelter and turned this into reproductive advantage.

Although many species modify their behaviour in synchrony with seasonal change, their temporal horizons are very limited compared with humans. As J. T. Fraser, the doyen among philosophers of time, has put it (Fraser, 1987):

> Many species can communicate their fears and plans, but except for a few examples of limited scope, they cannot generate or receive communication about the past. I can convey to a dog the idea 'I will feed you', and the dog will respond appropriately. But there is no way I could tell Fido, 'I've already fed you.'

Unlike the reindeer that migrated across the cold steppes of northern Europe during the Ice Age because their internal clocks dictated their movements, our ancestors came to choose, on the basis of their knowledge of the remembered past, when to ambush and massacre them.

We learned to cope, to manage and eventually to manipulate those changes to our advantage. But despite all our modern advances, our physiology and anatomy, formed over eight million or more years, are indelibly marked by the seasons to this day.

# 8

# TIMING REPRODUCTION IN HUMANS

In the Spring a fuller crimson comes upon the robin's breast;
In the Spring the wanton lapwing gets himself another crest;

In the Spring a livelier iris changes on the burnish'd dove;
In the Spring a young man's fancy lightly turns to thoughts of love.

ALFRED LORD TENNYSON, 'LOCKSLEY HALL' (1842)

In early Victorian England, when Tennyson was writing his thinly disguised poem about his love for a young woman, it was not just young men's fancy that turned in spring: conceptions peaked in late spring/early summer. Unbeknown to him, men's sperm quality is highest then, while *in vitro* fertilisation studies have suggested that the fertilisation and quality of the female embryos are highest during the spring and lowest in autumn (Vahidi *et al.*, 2004).

Diligent studies of parish records and local and national censuses have built up a picture of the seasonality of conception and birth in several northern European countries, including Sweden, Finland, England, Germany and Holland, dating back well into pre-industrial, pre-contraceptive pill and pre-elective Caesarean times. Although there is some variation between countries, births tended to peak around the spring equinox, followed by a secondary peak in September and a low during November and December.

This 'European' birth pattern is typical of agricultural populations at higher latitudes. It reflects a high frequency of conceptions in June and July and a low conception rate during the autumn harvest season (Lam & Miron, 1994). It is not by chance that June is still the most popular month in which to marry. The tradition began with the ancient Romans, when the month of June was named after Juno, goddess of women and marriage, who vowed to protect those who married in her month. A June or July conception and a subsequent early spring birth meant that the mother had recovered to some extent in time for the busy autumn harvest season in the following year.

Swedish records are particularly reliable, and in that country not only did this essentially pre-industrial pattern persist into the twentieth century rather than attenuating with increasing urbanisation and the corresponding rupture with the agricultural year, but in the years from 1969 to 1987 the peak-to-trough difference in monthly births was more than 30 per cent, nearly twice what it had been in the 1920s and 1930s. While Sweden clung tenaciously and unusually to the past, this is not so in Germany, where the birth peak has switched in the past 60 years from February and March to September (Lerchl et al., 1993).

The records in Louisiana in the southern USA also show a marked seasonality in birth. Unlike Sweden, the trough in the past has been in spring and the peak in autumn. This seasonal variation in Louisiana and in many of the states of the Union, even in higher latitudes, is difficult to explain in view of the seemingly reasonable explanations for the European pattern. In Louisiana the seasonal variance has been flattening in recent years, as the numbers of homes with air-conditioning has steadily increased. High temperatures in the southern states could have affected the sperm in various ways, or perhaps modified the ovulation characteristics of the female. Coital frequency may not have been maintained in the hot months. Or perhaps it was that the incidence of spontaneous abortion was higher and so the evidence of sexual activity did not show up nine months later?

The late Rick Condon, who tragically died on a field trip in the Arctic, and his co-author Richard Scaglion compared the birth seasonality of two societies in different hemispheres that they had individually studied in the 1970s. The Copper Inuit live in the Canadian Arctic, where the winter low is −30°C and the summer high is just over 7°C, accompanied by huge changes in the amount of daylight, wind speed and direction and, of course, ice conditions. The Samukundi Abelam live in Papua New Guinea, where the mean daily temperature seldom varies much from around 25°C and most days are relatively humid, although there is a wet season and a dry season (Condon & Scaglion, 1982). Although both societies had been in contact with Westerners, and the Inuits in particular were already beginning to live a very different life from their ancestors, Condon and Scaglion were still able to piece together the traditional patterns.

Samukundi life revolves around yams. The males tend the crop and their status is dependent on the size of the tubers they grow. The centrality of yams is associated with taboos and rituals. There is a six-month prohibition against sexual activity from July until the January harvest, and during this time there is a ban on sexual innuendo and joking and also contact with menstruating women. Births peak in October, and this is associated with the cultural sex taboo because there is no such seasonality among neighbouring tribes. The taboo is self-reinforcing because women who conceive soon after the harvest are in the third trimester and so are not sexually receptive during the critical yam-growing months of August and September.

Among the Inuit, 'social and economic patterns of winter concentration and summer dispersal are regulated by the change of seasons' (Condon & Scaglion, 1982). Winter is the time for socialising, and the dark months are spent in the settlement. As the weather improves in the spring, Inuit families go off on their own to camp outside the settlement and go ice-fishing and duck hunting. This enables them to have more privacy and intimacy. As a result the vast majority of conceptions take place in the spring and summer, and most births are in the first half of the year.

The rhythmicity in rainfall and humidity synchronises the yam-growing season for the Samukundi, and social behaviours and cultural taboos 'cement' this seasonality into their lives by the restrictions on sexual congress. This social rhythm synchronises the annual birth rhythms and imposes pronounced birth seasonality in the absence of any dramatic seasonal variation in the environment, including daylength. Among the Copper Inuit, climatic rhythms directly synchronise economic activities, social behaviours and physiological response, all of which, Condon and Scaglion conclude, contribute to the non-random distribution of births throughout the year.

In other societies, and perhaps unsurprisingly, straightforward biological resource factors also account for birth seasonality. Almost universal weight loss is a common feature of the pre-harvest season among the Lese women in the Ituri Forest of the Democratic Republic of the Congo. This weight loss is accompanied by lower levels of salivary progesterone and oestradiol, longer intermenstrual intervals, and shorter durations of menstrual bleeding. All these trends are reversed after the harvest as a positive energy balance is established. Over time this seasonal variation in ovarian function is reflected in a statistically significant seasonal pattern of conceptions after the harvest (Bailey et al., 1992).

Although seasonal timing of births might be expected in societies that live close to the natural world, it persists even in today's modern world. Despite all the contrivances we use to dominate nature and hide from seasonal change, a persistent seasonal variation with clear peaks and troughs in births is an almost universal characteristic of human populations, whether urban or rural, tropical, temperate or boreal. Quite why this seasonality still exists, even acknowledging that the seasonal peak and trough may vary, is unknown. It is surprising given that food availability is much more constant, and that our modern air-conditioned homes and offices and '24/7' lighting effectively mask us from the natural world. Despite its persistence, seasonality of birth is, however, far less marked in the industrialised nations than it used to be, being either virtually undetectable in some cases or of very low amplitude (about 5 per cent) and requiring large population statistics for detection (Roenneberg, 2004).

Humans are not seasonal breeders in the same sense as, say, sheep and hamsters. Both men and women are ready to procreate, more or less at the drop of a hat, more or less most of the time, provided that the female is neither pregnant nor lactating. So there is a puzzle as to how and why these largely opportunistic breeders even today can show birth seasonality.

It is not just conceptions and births that show seasonal variation. The frequency of sexual activity is seasonal, as are sexually transmitted diseases and the sale of contraceptives (Meriggiola et al., 1996). Although one would expect an interlinkage in timing between these activities, unpicking the various biological, economic, cultural and social factors that account for seasonality in human sexual and reproductive behaviour is not at all easy. During our history, the pressures on an agrarian economy that was largely dependent on the seasonal weather meant that social organisation, and with it conception and birth timing, was closely tied to the requirement to provide the necessary labour force for the peak times of planting and harvesting.

One possible biological explanation of birth seasonality is that it all began some five thousand or more generations ago, when early humans who were anatomically very similar to us started on the long trek out of Africa into the higher latitudes. These ancient hominins may have reproduced seasonally at the higher latitudes because of the variation in food availability. As in sheep, this could have been driven by a photoperiodic or even a circannual response.

However, there is disagreement over whether humans actually are, or ever were, photoperiodic or influenced by a circannual clock in any meaningful sense. Till Roenneberg of the University of Munich is of the view that we are photoperiodic and that human fertility and birth seasonality are largely susceptible to changes in daylength. With the late Jurgen Aschoff, Roenneberg exhaustively analysed the birth records of many countries. He explains how, in Spain, the seasonal conception rhythm that peaked in late spring and had proved relatively impervious to the global social changes that followed the Second World War changed markedly in the 1960s when Franco initiated a massive industrialisation campaign that included introducing extensive electrification and factories into rural areas. According to Roenneberg, industrialisation meant working indoors, and the reduced exposure to daylight modulated the photoperiodic response with the result that the conception peak shifted from spring and summer to autumn and winter. Combining photoperiod and temperature, which also affects the reproductive process in humans, accounts in Roenneberg's view, for much of the seasonality in human birth (Roenneberg, 2004) (Figure 8.1).

We ought to be photoperiodic because, as Tom Wehr, previously at the National Institute of Mental Health in Maryland, USA, has pointed out (Wehr, 2001):

Most of the anatomical and molecular substrates of the system that encodes changes in the duration of melatonin secretion and the receptor molecules that

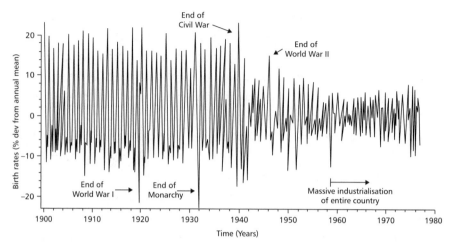

**Figure 8.1** Seasonal and social influences on human reproduction exemplified by the monthly birth rates in Spain from 1900 to 1978. Before 1940 the birth rhythm was highly regular, with a peak of conceptions in spring and annual differences of about 30 per cent (peak to trough). Social interventions, such as wars, introduce irregularities but have little effect on the overall rhythmicity. During and after the Second World War the rhythm's amplitude decreased but otherwise continued with the same characteristics. During the 1960s Franco launched a massive industrialisation campaign, introducing more extensive electrification and factories into many rural areas. Concurrently, the annual birth rhythms changed drastically. The amplitude declined even more and the phase of the conception peaks shifted to autumn and winter. (Redrawn from Roenneberg (2004).)

read this signal have been shown to be conserved in monkeys and humans, and the functions of this system appear to be intact from the level of the retina to the level of the melatonin-duration signal of change in season.... Epidemiological evidence suggests that inhibition of fertility by heat in men in summer contributes to seasonal variation in human reproduction at lower latitudes and that stimulation of fertility by the lengthening of the photoperiod in spring contributes to the variation at higher latitudes.

But Wehr adds the caveat, 'While photoperiodic seasonal breeding has been shown to occur in monkeys, it remains unclear whether photoperiod and mediation of photoperiod's effects by melatonin influence human reproduction.'

Studies on mammals have shown that the pattern of melatonin release depends on the daylength, as we discussed in Chapter 4. This signal provides mammals with a fix on the time of year. For humans, it was thought from the various studies that had been done that only those living at latitudes higher than 60° had a melatonin signal duration that depended on the photoperiod, and even then only slightly. However, when Helena Illnerová, working at the Institute of Physiology in Prague, which is at 50° N, studied a group of students in winter and summer, she found (Vondrasová *et al.*, 1997) that, under natural lighting conditions,

> exposure to an actual natural summer photoperiod from sunrise until sunset, phase advances the circadian oscillator controlling the morning melatonin decline and cortisol rise and shortens the melatonin signal duration as compared with the winter pattern.

Despite this, not everyone agrees that humans are generally photoperiodic, let alone that the photoperiodic response is the main factor in modern human birth seasonality. Among the sceptics is Frank Bronson of the University of Texas, who carries considerable weight as one of the world's leading authorities on reproduction. In a paper provocatively titled 'Are humans seasonally photoperiodic?' (Bronson, 2004) he considers that the findings of the various demographic studies are inconsistent in that it is difficult, if not impossible, to discern an overall pattern across countries and climates that reduces the issue to that of one or two major causal factors. He considers that the differences in seasonal reproduction among humans around the globe reflects the variation in photoresponsiveness between individuals, some of whom he accepts do respond to changes in photoperiod, whereas others do not, and there are many in between with varying degrees of responsiveness that may be manifested as a variation in critical daylengths. It is an argument that may seem at first sight somewhat pedantic, but it centres on the important distinction between the epidemiological approach, with its emphasis on population averages, and one based on individual variation.

Bronson's argument is that among our early ancestors, those living in the latitude belt 10–15° either side of the equator were probably not responsive to photoperiod. Michael Menaker, another of the doyens of biological rhythm research, is another sceptic. In answer to the question 'If seasonal rhythms are so robust in other animals, and if humans have the equipment to express them, then how did we ever lose them?' Menaker replied, 'What makes you think we ever had them? We evolved in the tropics' (personal communication).

This does not mean that seasonal change does not occur at all in the tropics. But it is true that there is a far less marked need to time reproduction to a particular time of the year, whereas there is an obvious and marked adaptive value in regulating reproduction at higher latitudes so that birth occurs at the optimal time of seasonal food availability. Whether or not we became similar as we moved into the higher latitudes is a moot point, but as our ancestors gained greater control of their environment over the millennia, seasons probably became a less significant evolutionary force (Wright, 2002).

In Bronson's view, responsiveness to photoperiod has been highly variable among humans since early in our evolutionary history. He argues that over the past few generations, global migration patterns among humans would have increased this heterogeneity as people indigenous to the tropics have moved to higher latitudes and to some extent vice versa, while the supplementation of short winter days with artificial light would have masked the photoperiodic effect among those in the general population who were responsive.

The upshot of all this is that even among responsive individuals, the endogenous seasonal rhythm that may have existed among our ancestors living in higher latitudes has been weakened. Combined with the masking effect of artificial light and heating and the variation in critical daylength among responders, this suggests to Bronson that if everything is taken into account, a population is a mix of responders and non-responders in varying proportions and that these admixed populations are the basis for the inconsistencies in the data that show up when the attributes of a heterogeneous population are averaged.

Which may or may not be the case? The question still comes back to whether we generally are or are not photoperiodic. Roenneberg asks whether, if we went back to a way of life circa 1930, still dominated by natural rhythms of light and dark, the annual pattern of conceptions would be similar to that in 1930; in other words, our endogenous seasonal rhythms would manifest themselves. An analogous event occurs in the mould *Neurospora crassa*, strains of which have been kept in artificial laboratory conditions for decades on end yet can still show strong photoperiodic responsiveness even though there have been generations that have had these rhythms effectively masked by the laboratory environment. But we are not *Neurospora* and we are not going to be able to do the experiment. Although Illnerová showed that a melatonin response is possible in humans under natural conditions of bright light, this is not the same as showing that human birth seasonality is synchronised through a seasonal photoperiodic mechanism.

Temperature may well account for the striking difference between North Americans and western/central Europeans in seasonal conception rates. The former show clear bimodal patterns with two peaks per year, and the latter a unimodal pattern with one dominant annual peak around the spring equinox and a second minor 'blip' in autumn. The bimodal pattern of conception, typical for North America, also exists in eastern Europe and this suggests a climatic influence, specifically of temperature. Very hot summers and cold winters are generally found in continental regions (such as North America or eastern Europe), whereas milder climates with less severe seasonal differences are found in more coastal regions (western and central Europe) (Foster & Roenneberg, 2008).

In countries that never experience high temperatures (for example in the Arctic and near-Arctic), conception maxima generally correlate with the time of highest annual temperatures. By contrast, in those predominantly equatorial and near-equatorial countries that never experience really low temperatures, conception correlates with the annual lowest temperature (Roenneberg & Aschoff, 1990). There seems to be a negative correlation between conception and temperature in hot locations, whereas this correlation is positive in cold environments. In regions between these extremes, conceptions are always highest when morning minimum temperatures reach approximately 12°C. A clear latitudinal gradient exists in the timing of human reproduction from the poles to the equator, gradually changing from later to earlier. This gradient could well be related to temperature (as it is for the clines in plant and tree flowering).

If not photoperiod and/or temperature, what else could cause the excesses in birth seasonality that have been so well documented? Is it food availability patterns, sexual behaviour, contraception, holiday timing (the September secondary peak in births is a function of conception at Christmas or New Year– there is a holiday effect even in non-Christian countries such as India and Israel) or agricultural cycles? Is it a combination of these or are there a few causal factors that are modulated by minor factors to provide the overall patterns that we see around the world?

Gabrielle Doblhammer of the Max Planck Institute for Demographic Research thinks that although neither the weather nor photoperiodism alone causes birth seasonality, they have a role to play in that 'climate and photoperiod appear to explain shifts in the amplitude of the peaks and troughs across region, but not the fundamental presence of the peaks and troughs in the first place' (Doblhammer *et al.*, 2000).

She is still searching for what she calls a complete statement of a 'unifying theory of human birth seasonality.' On the basis of a detailed historical study of Austrian birth patterns, she suggests the 'resilience hypothesis', in which a multiplicity of small causes slightly shift, dampen and enhance the patterns, which are themselves caused by a very few resilient and much stronger causes. She and her co-workers note (Doblhammer *et al.*, 2000):

> Previous research has been effective in identifying critical factors related to the amplitude of the three major seasonal features in human birth rates around the world [geography, climate, and agricultural practices – Ed]. However, specification of the fundamental causes of those three features in the first place is still an ongoing and fascinating research endeavour.

In one intriguing study of the records of Dutch women who married between 1802 and 1929 and subsequently had children (a research population that may have experienced more natural patterns of light and temperature than modern populations brought up in an environment of artificial lighting and heating), those born at certain times of the year seemed to be more likely to show a seasonal variation in their rate of conception than other women (Smits *et al.*, 1997). Quite why the fertility of these women was inhibited at certain times of the year is a mystery. One possible explanation is that there is a melatonin-dependent seasonal variability of the quality of the oocyte, the cell that develops into the ovum (egg).

One reason for the difficulty in coming up with an explanation is the eccentric duration of the human reproductive process. If there were such a thing as a designer, intelligent or otherwise, there would not be a nine-month gestation in humans in the higher latitudes, but a one-year period, which would give the best fit between optimal mating time and optimal birth time. In an ideal world of hunter-gatherers, one imagines that humans would tend to conceive in mid to late spring when overall environmental conditions were most favourable, and births would occur a year later when there would be a maximal food supply for the infant and mother.

The nine-month gestation period of humans simply does not add up. A nine-month gestation means that in temperate zones, April or May mating would lead to midwinter births, whereas if April or May were the optimal time for births, as it is with many other species, a nine-month gestation period would imply early autumnal mating, which is not the season that Tennyson suggested.

But nine months it is for humans, and it is a similar period for our primate chimpanzee and gorilla cousins living in the mildly seasonal tropical rainforests of Africa. This suggests that the 270 or so days for human gestation is a throwback to well before the time when our ancestors started their trek away from the tropics.

The length of the gestation period is largely but not entirely a matter of size, or rather scale – in mammals at any rate. Large-bodied species have big offspring that take longer to develop, and development is also prolonged for offspring that are born at an advanced stage of development. Elephants take about 640 days, whereas in chimpanzees and gorillas it is between 240 and 270 days. In cats and dogs it is a mere 60 days, in rabbits just 33 and in mice 20. Even among the species of apes and monkeys, gestation period seems to be a matter of size. For rhesus monkeys it is 164 days and for baboons it is 187 days.

We are stuck with about 270 days, give or take a week or so. Birth timing is probably triggered when the energy input from the maternal circulation is exceeded by the energy demand of the foetus – this is one reason why twins tend to be born prematurely. Babies are born when the foetus is beginning to starve in the womb. Hence the gestation period is fairly inflexible and more or less impervious to proximal factors in the environment. If the external energy resources are low, gestation stays at around 270 days, and the result is a low-birth-weight baby, which implies a higher risk of mortality.

Those of us living in those higher latitudes have just had to make the best of this nine-month timing and so we used a variety of social and cultural as well as biological means to try to align birth time to the optimal time of year.

It may be that present birth seasonality is an echo of previous human generations when birth month had a selective effect on fertility and overall evolutionary fitness. In northern temperate latitudes, men born in autumn have fewer offspring and a higher probability of remaining childless than men born in spring. Among women, those born in summer months have fewer children than those born at other times of the year (Huber *et al.*, 2004).

Virpi Lummaa and Marc Tremblay investigated the effect that the conditions experienced during early development have on overall reproductive fitness by carefully analysing the records of Canadian women living in a tough climate in the Saguenay region on the north shore of the St Lawrence River in Quebec, Canada (about 48° N), in the nineteenth century. There is a vegetative season from mid-May to mid-September, but frost at the end of May and at the beginning of September is not unusual. Lummaa and Tremblay looked at the cohort born between 1850 and 1879, including the survival and complete reproductive history of their surviving offspring (born between 1866 and 1926) who married in the same population (Lummaa & Tremblay, 2003). The female population they studied was almost entirely French speaking, Catholic, mainly agricultural and culturally homogeneous.

Remarkably, they found that the month of birth, which is a useful surrogate for the conditions experienced during early development, predicted a woman's 'fitness' measured by the number of grandchildren produced. Women born in June (the 'best' month) had on average at least seven more grandchildren than those born in October (the 'worst' month). Those born in high-success months got married earlier (if under 30 years old), gave birth to their last child later, had longer reproductive lifespans, gave birth to more children and raised more to their own child-bearing age.

But was it the timing of birth of these Canadian women that was the crucial factor, or was it the timing of conception? Peter Ellison, who heads Harvard's Reproductive Ecology Laboratory, argues in favour of the latter (Ellison, 2003):

> The ability of a woman to divert metabolic energy towards reproduction in the early gestation period is a significant factor in the success of that pregnancy. Indeed, even though the direct energetic costs of the embryo and placenta are minimal in the first few months of gestation, significant fat accumulation occurs during this period. These stores are later drawn upon to meet the high costs of late gestation and early lactation. The efficiency of fat storage is increased in early pregnancy in direct proportion to oestrogen levels. Successful reproduction depends on a continuous diversion of energy from a woman's current metabolic budget. It is the potential for sustaining an ongoing investment that matters most.

The driving force in all evolutionary activity is the production of young, who in turn produce more young. Maximising reproductive fitness is the *sine qua non* to understanding all biological activity. In theory, for the mother to maximise reproductive fitness she could extend her reproductive lifespan, reduce the recovery time before the next conception to as short a duration as possible, and increase the reproductive rate of her offspring. The constraints include the biological demands of child rearing, which limit, for example, multiple births at any one time and also the time needed to rear an infant until it is weaned.

Primates in general are slow breeders. At the equator, primate conception is largely opportunistic. The mother conceives, gestation proceeds, and lactation continues after birth. This conception–gestation–birth–weaning cycle is largely fixed in duration. A female gorilla spends about four years raising her infant before she prepares to conceive again. When lactation stops there is then a period of recovery before the mother can once again conceive. The duration of this recovery process will be determined by how quickly she can re-adjust her metabolic state so as to be in the best condition for the next conception–gestation–birth–weaning cycle. As food is more or less a constant at the equator, this recovery time will also be a fairly fixed period.

The male primate, in contrast, is usually always ready to mate. Males invest nothing like as much energy in reproduction as females, but they also face trade-offs. Although it is an optimising strategy for males to be always ready or near-ready to copulate, there is a higher mortality among males in such a permanently aroused state because they are also in a continual state of aggressive behaviour towards other males, which can get them into difficulties – in other words, dead.

What all this amounts to is that there is an optimum interval between births for the female primate at the equator. This results in a periodicity in the production of young for an individual that is largely independent of the time of year, and at the population level there is little or no seasonality to conception and consequent birth.

When our hominin ancestors moved to higher latitudes they took this equatorial pattern of reproduction with them. But at the higher latitudes the seasons and their effect on food availability mattered critically to the success or otherwise of a pregnancy and also to the development of the young. And if this were not enough, there is also the problem that, during human development, the infant is subject to various seasonally based events that are separate from food availability. For instance, Doblhammer cites historical studies that, admittedly based on the recent past, outline challenges that resonate with the situation confronting our ancestors 50,000 years ago (Doblhammer, 1999):

> In Italy the summer cohorts were advantaged because they went through the summer with the full protection afforded by breast-feeding and reached winter at an age when they were less vulnerable to viral infections of the respiratory tract. The winter cohorts, on the other hand, were exposed to the impact of the cold season on respiratory diseases in their first months of life. This was then followed by the hot summers and the accompanying viral infections of the digestive tract, at a time when the protection of breast-feeding was diminished.

To deal with this, the hominins had either to somehow align the optimal conception–birth interval seen at the equator to what would be the time of year that provided for the 'best' overall reproductive success in the local conditions, or to evolve a photoperiodic response to regulate conception and birth to the same effect.

Because it seems at the very least problematic as to whether humans have ever been actively photoperiodic, the alternative strategy of linking female metabolic state to conception, so as to best cope with gestation, is attractive. This would allow for a late spring or early summer peak in conception, when the food availability and female metabolic state would have been at their highest, and a subsequent birth peak around the spring equinox. In this scenario, food availability drives maternal metabolic state, which in turn drives successful conception and gestation.

Such a strategy could also adapt to local variations. As the connections were made between procreation, conception (perhaps through absence of menstruation), pregnancy and birth, so a cultural overlay would have been added that helped to define conception timing via fertility rituals and the like. In the same way, in modern times Danish women who were questioned in a survey are reported to have a desire to conceive in summer so as to give birth in the following spring, as they believe that spring babies have a better health outcome (Basso *et al.*, 1995); of course they have modern contraception to enable them to choose the timing of conception.

When it comes to the timing of reproduction, humans are not like sheep. Sheep can only give birth at a particular time of year when food availability can sustain milk production and weaning of the lambs. It may be that Ellison's view that it is not the resource availability at birth that is the critical factor for humans, but the metabolic condition of the mother at conception, is correct and that a woman has a conception–gestation–birth cycle that is attuned to her condition at the time of conception. Whether it applies universally remains unclear owing to a paucity of detailed longitudinal studies in other human societies. Because human offspring have a period of prolonged dependence on the mother and the energy demands of both pregnancy and lactation are so high and are spread over such a long period, there may be no optimal time of year for conception based on food availability alone. Rather, the mother has to be able to sustain good nutritional health across multiple seasons. Seasonal reproduction in humans is likely to be complex and dependent on multiple environmental and biological variables, and as a consequence it may well have very variable timing and amplitude across different societies and living conditions.

This is not to deny that the timing of the birth itself has a profound and long-term effect. Studies of Gambian villagers who experience marked annual cycles in food availability and disease have demonstrated a strong correlation between birth season and premature death. In this African population, young adults who die of infectious diseases are much more likely to have been born during the 'hungry season' than at more favourable times of the year (Moore *et al.*, 1999).

Gestation events affect not only the immediate offspring but also the subsequent generation, as was spectacularly demonstrated in women who became mothers during or soon after the Dutch famine of 1944–1945. About 30,000 Dutch people starved to death during the *Hongerwinter*. There was a 300-gram decrease in mean birth weight of the babies of women who had conceived at the start of the famine and so had endured poor nutrition throughout the pregnancy. These low-birth-weight babies did not suffer from adverse effects on their subsequent fecundity in adulthood, but they were themselves more likely to give birth to offspring of reduced birth weight. This reduced birth weight in the second generation was associated with a higher frequency of stillbirths and early infant mortality. The result was that females exposed to the famine *in utero* had reduced reproductive success compared with those who were born before or after the famine (Lumey & Stein, 1997).

What and how much pregnant mothers eat can profoundly affect the development of their offspring. Although the genome lays out the basic plan for the brain and central nervous system, the actual development of an individual *in utero* and *ex utero* occurs while it is growing in a complex environment. In temperate zones, the external environment varies with the seasons and affects the maternal environment in such a way as to affect early development, causing potential downstream effects. Even identical twins are not actually identical!

In the Dutch famine, the foetuses were growing in the womb of nutritionally deprived mothers and so the foetus was under nutritional stress. Detailed studies in adult life of babies born during the famine found that maternal nutrition during early gestation can permanently influence the lipid profile in later life. Exposure to famine in early gestation led to a higher ratio of LDL to HDL cholesterol (known popularly as 'bad' and 'good' cholesterol) in adult life. Other studies in humans confirm that maternal nutritional intake during pregnancy can have permanent effects on health in later life (Roseboom *et al.*, 2000).

It has been suggested that, in response to the nutritional stress, the foetus adopted a 'thrifty' metabolism which it kept throughout its adult life, even though the nutritional stress was relieved in early infancy. When the next generation was conceived, the maternal metabolism was still 'thrifty' and so the foetus received fewer nutrients than required, so it in turn developed a 'thrifty' metabolism (Sapolsky, 2004). This interaction between the genome and the environment is the basis of the discipline of epigenetics. The manner in which the genome is decoded is intimately connected with the environment at the time. For example, the switching on and off of genes is modulated by the histone–protein complex that immediately surrounds the DNA. There is a growing tendency to refer to the epigenome rather than the genome as such when discussing development.

Researchers have linked exposure to severe hunger in pregnant women to a range of developmental and adult disorders in the offspring, including low birth weight, diabetes, obesity, coronary heart disease, and breast and other cancers, and at least one group has also associated exposure with the birth of smaller-than-normal grandchildren (Pray, 2004). This line of thinking has become consolidated into the notion of biological plasticity. Essentially, organisms, including ourselves, that have had to deal with seasonal change (and the attendant consequences in terms of food and water supply, exposure to pathogens, temperature and thermoregulation), social grouping and all the environmental consequences of annual change in daylength have evolved so that there is considerable flexibility during the developmental stages when the genomic framework contained within the DNA is translated into the actuality or phenotype of the offspring. The offspring develops anticipating the conditions it expects to meet during its life course.

If the conditions it meets during pregnancy are aberrant – say, a temporary famine in western Europe – then the infant will have a lifelong metabolism geared to scarcity even though it will live in a metabolically rich environment. There are real fears that in our modern society the external metabolic environment that the foetus senses through the exchange with the mother's placenta may be a false prediction of the future because such factors as smoking can affect nutrient supply; alternatively, the mother may eat an unbalanced diet with similar effect. The result is that we tend to be born with a metabolism geared to the thrifty end of a spectrum but we live in a nutritionally dense world. The result is that we eat more than we should and more than we can handle – and we get fat (Gluckman & Hanson, 2006)! This interaction of genome and maternal environment has complex and surprising effects on the subsequent health of the offspring.

# 9

# BIRTH MONTH EFFECTS

The Child is father of the Man;

WILLIAM WORDSWORTH, 'THE RAINBOW' (1807)

While there may be a best time for human conception, predicated to the maternal nutritional state, is there an ideal month for the birth of human babies? It seems absurd that the month in which you are born can affect your future life chances, but how long you live, how tall you are, how well you do at school and college, your body-mass index as an adult, your blood pressure, the age of menarche for a girl and menopause for a woman, the likelihood of an eating disorder, your fecundity, your likelihood of autism or of panic disorder, your morning versus evening preference and how likely you are to develop a range of diseases, including devastating conditions such as schizophrenia, are all correlated to some extent by the time of year in which you emerged from the womb. Winter-borns show increased novelty seeking and sensation seeking relative to those born during the remainder of the year (Eisenberg *et al.*, 2007). There have even been claims from studies of American baseball players that the month of birth even affects whether you are more or less likely to be left or right-handed (Abel & Kruger, 2004).

Although this seeming predestination has been the happy hunting ground of astrologers, fortune tellers and seekers, and a raft of pseudo-scientific inquirers, one of the first people to look at it seriously was Ellsworth

Huntington, an American geographer and professor of economics at Yale and, less comfortingly, a leading eugenicist.

A recent and very careful study of more than two million Danes and Austrians who died in the last three decades of the twentieth century suggests that, in these two countries at least, lifespan and month of birth are related (Doblhammer, 1999). The studies excluded deaths under 50 years of age and were carefully controlled for possible confounding effects such as the seasonal distribution of death, which is higher in winter than in summer. Allowing for all this, in Austria, for all major groups of causes of death including accidents and suicides, mean age at death of those born in the second quarter (April to June) is significantly lower than that of individuals born in the fourth quarter (October to December).

In this context, 'significantly' is a statistical term meaning that the effect is real rather than necessarily being big. As it happens, the lifespan of people born in the second quarter is about 101 days *below* average, whereas the lifespan of those born in the fourth quarter is 115 days *above* average.

Among Danes, those born in the second quarter die about 60 days *below* the average age, whereas those born in the last quarter die about 47 days *above* the average age. If you are Austrian, then on average those born in the fourth quarter live a little over 200 days longer than those born in the second quarter, whereas in Denmark the difference is about 100 days (Doblhammer, 1999). For most of us that would be a difference worth having.

Gabrielle Doblhammer studied lifespan in the USA, and her investigation concluded that those born in the autumn live about 160 days longer than those born in spring. She also found a significant month-of-birth pattern for all major causes of death, including cardiovascular disease, malignant neoplasms – in particular lung cancer – and other natural diseases such as chronic obstructive lung disease (Doblhammer & Vaupel, 2001).

In the southern hemisphere, the pattern is shifted by half a year. The mean age at death of people born in Australia in the second quarter of the year is 78 years; those born in the fourth quarter die 125 days earlier. Furthermore, the lifespan pattern of British immigrants to Australia is similar to that of Austrians and Danes, and is significantly different from that of Australians (Doblhammer & Vaupel, 2001).

If birth in October to December is related to increased longevity in the northern hemisphere, an obvious question is why are we all not born in the last quarter of the year rather than in the peak birth months of March or April. Apart from the issue of maternal nutrition at conception perhaps being the determining factor in birth timing, longevity in itself is not a basis for evolutionary selection – natural selection is not interested in us after we have finished child-bearing. There may be a possible longevity advantage in that grandparental assistance in child rearing may improve consequent offspring survival, but that is highly unlikely to be anywhere near sufficient to offset other factors.

It is not just length of life, but also disease, that seems to be determined by date of birth. Long-term studies in Sweden have shown that infants born in the summer months are more likely to develop diabetes mellitus before they reach the age of 15 than those born in the other seasons. There are many other factors that could influence susceptibility to disease, but as Allan Smith of the University of California at Los Angeles commented (Spears, 2004):

> Right now, we don't know what makes some people develop diabetes, for example. Some of the reason is in your genes, but not all. The seasons seem to play some role, too. So there's an association and if we ever want to cure [diabetes], or just treat it better, we have to know what goes wrong. The point of this knowledge is to understand what causes these diseases in the first place.

Similarly, summer babies have a greater than average chance of developing coeliac disease, a digestive disorder. The actual cause of that problem, however, is unknown. Emmanuel Mignot, professor of psychiatry at Stanford University, who discovered an increase in narcolepsy among people born in March, has pointed out 'Environmental factors are very difficult to study and are very speculative, it's nearly impossible to find out what could be involved, like finding a needle in a haystack' (Spears, 2004).

But Smith's argument is crucial, and even if the differences related to birth month are not large, they can still be of enormous importance in helping us to understand the mechanisms that can lead to disease.

John McGrath of the University of Queensland reinforces this point when he talks about why seasonal effects are so useful in research. Although he is talking about schizophrenia, he is essentially generalising when he says (Swan, 2005):

> It's very exciting times to be doing schizophrenia research because the old flat earth policy that schizophrenia has no gradients is very depressing [be]cause you need gradients to get traction to generate candidates. ... if you're born in winter and spring you've got an increased risk. If you're a male you have an increased risk so this is the stuff that's the grist for the mill. We are so ignorant ... it's like fever, the ancient physicians used to think fever was one illness [and] that's about where we are now. Psychosis is a final common pathway for the broken brain and we have to tease apart all the myriad of factors that contribute to that.

This idea that gradients provide a handle to get at some of the root causes of illness is not new, it is just becoming far more sophisticated. The percentage differences of a gradient may be small, but they are real and most probably reflect some underlying mechanism and not blind chance.

Traction comes from many angles, but in epidemiology it often starts with gradients. Why do babies born in central Europe in the second half of the year outlive those born in the first? Why are babies born in the USA in March more likely to develop narcolepsy in later life than those born in other months? Why are some mental health disorders more prevalent in those born in certain months?

In the northern hemisphere, babies born in the early part of the year are 6–8 per cent more likely than others to develop schizophrenia in later life (Torrey et al., 1997). An observation from epidemiological analysis poses the questions: Why is there this gradient? What is it that can have this effect?

Schizophrenia is a devastating illness. It is a severe mental illness characterised by persistent defects in the perception or expression of reality. A person experiencing untreated schizophrenia typically demonstrates grossly disorganised thinking and may also experience delusions or auditory hallucinations. Although the illness primarily affects cognition, it can also contribute to chronic problems with behaviour or emotions.

In the UK there are about 250–300 adults in every 100,000 with schizophrenia at any given time. In the USA, more than two million people have been diagnosed with a schizophrenic disorder. The illness wrecks lives and families. One out of every ten people with schizophrenia eventually commits suicide.

People who develop schizophrenia in Europe and North America are more likely to have been born in the winter and early spring (February and March in the northern hemisphere). The subjects who were born during these months had a slightly higher than average rate of schizophrenia, whereas subjects born in August and September had a slightly lower than average rate. There seems to be about a 10 per cent difference in risk of schizophrenia between the higher-risk (winter and spring) and lower-risk months of birth (Castrogiovanni et al., 1998).

A word of caution is needed here. Statistical assertions based on large population studies tell us nothing about any individual. The vast majority of individuals with these diseases are not born during the months of excess births, and most individuals born during these months do not develop schizophrenia or bipolar disorder. Although a 6–8 per cent variance sounds a lot, it is very small in terms of its impact on the overall incidence of these diseases.

Could the findings in schizophrenia and other pathologies outlined in Figure 9.1 represent anything more than a statistical artefact? For example, in schizophrenia, the condition is age linked. Because somebody born in the first quarter of the year is older than somebody born in the last quarter of the same calendar year, this age-lag has been suggested as a possible explanation for a birth month effect in a given calendar year. But when this age effect has been corrected for, the seasonal effect still stands. Another possibility is that because there are more births overall in spring, the rhythm in schizophrenia results from the same causes (Torrey et al., 1997). This seems unlikely, however, because the pattern of conception and birth differs between North America and Europe, but the peak birth months for schizophrenia are the same for both regions (late winter and early spring). So the general consensus is that for conditions such as schizophrenia, and for other pathologies, there is an influence of conception, pregnancy or birth timing and that the results do not represent a statistical artefact (Figure 9.1).

There are many putative causes for schizophrenia, and Matt Ridley ran through some of them in *Nature via Nurture* (Ridley, 2004). But he was

| Condition | Jan | Feb | Mar | April | May | June | July | Aug | Sept | Oct | Nov | Dec |
|---|---|---|---|---|---|---|---|---|---|---|---|---|
| *General Pathologies* | | | | | | | | | | | | |
| Asthma (UK) | | | | | | | | | | ▓ | ▓ | |
| Asthma (UK) | | | | | ▓ | ▓ | ▓ | ▓ | | | | |
| Asthma (Denmark) | | | ▓ | | | | | | | | | |
| Crohn's disease (Israel) | ▓ | | | | | | | | | | | ▓ |
| Childhood Diabetes mellitus | | | ▓ | ▓ | | | | | | | | |
| Glaucoma | | | | | ▓ | ▓ | ▓ | | | | | |
| Hodgkin's disease | | | ▓ | ▓ | | | | | | | | |
| *Psychiatric Disorders* | | | | | | | | | | | | |
| Alcohol abuse | | | | | | ▓ | ▓ | | | | | |
| Autism | | | ▓ | | | | ▓ | | | | | |
| Bipolar disorder | ▓ | ▓ | | | | | | | | | | |
| Eating disorder | | | | ▓ | ▓ | ▓ | | | | | | |
| Personality disorder | | | ▓ | ▓ | ▓ | | | | | | | |
| Neuroses | ▓ | ▓ | | | | | | | | | | |
| SAD | | | ▓ | ▓ | ▓ | ▓ | | | | | | |
| Schizoaffective disorder | ▓ | ▓ | | | | | | | | | | |
| Schizophrenia (N. hemisphere) | ▓ | ▓ | | | | | | | | | | ▓ |
| Schizophrenia (S. hemisphere) | | | | | | ▓ | ▓ | | | | | |
| Suicidal behaviour (W. Australia) | | | | | | | | ▓ | ▓ | | | |
| *Neurological Illness* | | | | | | | | | | | | |
| Alzheimer's disease | ▓ | ▓ | | | | | | | | | | |
| Amyotrophic lateral sclerosis | | | | ▓ | ▓ | | | | | | | |
| Down syndrome | | | | | ▓ | ▓ | | | | | | |
| Epilepsy | ▓ | ▓ | | | | | | | | | | ▓ |
| Mental retardation | | | | ▓ | ▓ | | | | | | | |
| Motor neuron disease | | | | ▓ | ▓ | | | | | | | |
| MS (Northern hemisphere) | | | | ▓ | ▓ | | | | | | | |
| MS (Southern hemisphere) | | | | | | | | | | ▓ | ▓ | |
| Narcolepsy | | | ▓ | ▓ | | | | | | | | |
| Parkinson's disease | | | | ▓ | | ▓ | | | | | | |

Month of <u>Birth</u>

**Figure 9.1.** The relationship between birth month and increased incidence of disease in later life. Considerable care has to be exercised in accepting this list as definitive; for example, compare the differing results associated with birth month and asthma. References: asthma (UK) (Anderson *et al.*, 1981; Smith & Springett, 1979), asthma (Denmark) (Pedersen & Weeke, 1983), Crohn's disease, (Chowers *et al.*, 2004), childhood diabetes mellitus (Rothwell *et al.*, 1996), glaucoma (Weale, 1993), Hodgkin's disease (Langagergaard *et al.*, 2003), alcohol abuse (Castrogiovanni *et al.*, 1998), autism (Castrogiovanni *et al.*, 1998), bipolar disorder (Castrogiovanni *et al.*, 1998), eating disorder (Castrogiovanni *et al.*, 1998), personality disorder (Castrogiovanni *et al.*, 1998), neuroses (Castrogiovanni *et al.*, 1998), seasonal affective disorder (Battle *et al.*, 1999; Castrogiovanni *et al.*, 1998), schizoaffective disorder (Castrogiovanni *et al.*, 1998), schizophrenia (northern hemisphere) (Battle *et al.*, 1999; Castrogiovanni *et al.*, 1998; Hafner *et al.*, 1987; Hare & Moran, 1981; Hare *et al.*, 1974), schizophrenia (southern hemisphere), (Hare & Moran, 1981; Hare *et al.*, 1974), suicidal behaviour (W. Australia) (Rock *et al.*, 2006), Alzheimer's disease (Castrogiovanni *et al.*, 1998), amyotrophic lateral sclerosis (Torrey *et al.*, 2000), Down syndrome, (Castrogiovanni *et al.*, 1998), epilepsy (Torrey *et al.*, 2000), learning disability (Castrogiovanni *et al.*, 1998), motor neuron disease (Castrogiovanni *et al.*, 1998), multiple sclerosis (northern hemisphere) (Battle *et al.*, 1999; Sadovnick *et al.*, 2007; Torrey *et al.*, 2000; Willer *et al.*, 2005), multiple sclerosis (southern hemisphere) (Willer *et al.*, 2005), narcolepsy (Dauvilliers *et al.*, 2003), Parkinson's disease (Battle *et al.*, 1999; Castrogiovanni *et al.*, 1998; Torrey *et al.*, 2000).

mainly interested in showing that the old Freudian-based idea that it was all the fault of the unloving mother was just so much guilt-inducing hogwash. At the last count there were at least ten ideas as to why it should be sensitive to birth month.

Some are more likely than others. Attempts to explain the seasonal difference include an appeal to the notion that schizophrenia tends to run in families and hence the suggestion that 'in the summer, people wear fewer clothes in bed, and ... a schizoid spouse is more likely then to notice his (or her) co-spouse there and accordingly to initiate sexual behaviour' (Ridley, 2004).

Far more plausible is the idea that the difference in the amount and intensity of light in the different seasons could have an effect. John McGrath has shown that vitamin D deficiency in pregnant animals leads to developmental brain abnormalities (McGrath, 1999). He has pointed out that, with regard to schizophrenia in humans, 'Our own theory ... which is highly speculative ... is that it's due to low prenatal vitamin D' (Swan, 2005). Vitamin D requires sunlight for its production in the body, and the argument runs that when people moved from the countryside to cities during industrialisation, the overcrowding into tenements and the soot and smoke meant that many of the inhabitants and children in particular received less natural light; we know this because of the emergence of rickets, which is caused by vitamin D deficiency. McGrath accepts that the jury is still out on the validity of his idea.

For the past two decades there has been something of an academic brouhaha focused on the role of infectious agents and viruses in particular in schizophrenia. Influenza comes in many forms, but some types at least have a marked seasonal occurrence and it has been linked in many studies to an excess number of winter births culminating in schizophrenia. In 2004, Ezra Susser and his team at Columbia University provided direct evidence that influenza is indeed a culprit (Brown *et al.*, 2004). They identified antibodies against the influenza virus in frozen serum samples taken from mothers who were pregnant during a 1960s child health development study in California, and were able to show that actual exposure to the influenza virus during the first trimester of pregnancy increased sevenfold the offspring's chances of developing a major mental illness, including schizophrenia. If the exposure occurred slightly later, between the midpoint of the first trimester and the midpoint of the second trimester, the risk was still increased, but the overall effect was a threefold increase. Exposure to the influenza virus even later in the pregnancy did not increase the risk to the offspring. What makes this study so important is that although it was based on a small sample and so its results were not statistically significant, it measured actual exposure to influenza rather than relying, as other studies had done, on correlating birth dates with known influenza outbreaks or on maternal recall.

It must be remembered that the overall risk is still small. About 97 per cent of children born to women who get influenza in the early to mid stages of pregnancy will not develop schizophrenia. But some 14 per cent of cases of schizophrenia might not have occurred had exposure to influenza during that period been prevented. One obvious suggestion, if the findings turn out to be valid, is that all women of childbearing age should be vaccinated against influenza. Why not just pregnant women? The worry here is that because the mechanisms underlying the connection between schizophrenia and influenza are not known, there could be harmful side-effects on the foetus if women were vaccinated during pregnancy.

This potential interaction has furthered speculation about a 'two-hit hypothesis', in which schizophrenia or bipolar disorder might be predisposed by a seasonal factor occurring during the perinatal period and then precipitated by another factor, not necessarily seasonal, such as cannabis, many years later.

The incidence of a stack of non-trivial behaviours, physical attributes and diseases are related to birth month. Among 180,000 patients who died of cancer in the USA, the average lifespan of those born in winter was 1.5 years longer than that of those born in summer. In Sweden, women born in June have a 5 per cent higher risk of developing breast cancer than women born in December (Kristoffersen & Hartveit, 2000).

At the beginning of this chapter we listed some of the characteristics that seem to demonstrate a season of birth effect. At the risk of repetition, everything from shyness in children, onset of menarche, novelty seeking in adolescence, owl or lark proclivity in adults, height and blood pressure in adult life, lifespan, and the incidence of multiple sclerosis and neurological disorders such as epilepsy have been claimed to be related to birth month with varying degrees of robustness in the studies. The list in Figure 9.1 shows the relationship between birth month(s) and enhanced incidence of disease in later life. Because the data come from different countries and different studies, care has to be exercised in accepting this list as definitive. Furthermore, as seen for asthma, results from a single country, let alone between countries, can be somewhat contradictory and await more data and analysis.

In terms of the best month in which to give birth in this context, 'you pays your money and you takes your choice'. Each month would seem to have long-term consequences. And the importance of this effect cannot be overstated. For example, the incidence of multiple sclerosis varies with latitude. It is virtually non-existent near the equator. There is a convincing latitudinal gradient in Australia, where the risk in temperate Tasmania is fivefold that in subtropical Queensland. George Ebers of Oxford University has coordinated a study that looked at more than 40,000 Canadian, British, Danish and Swedish patients with multiple sclerosis. His team's research has provided impressive evidence that in the northern hemisphere significantly fewer people with multiple sclerosis were born in November, which was the month of lowest later incidence, and significantly more were born in May. There was a 13 per cent increase in risk of multiple sclerosis for those born in May compared with those born in November.

In the southern hemisphere, the situation is reversed, and November is the peak month and May the lowest. Ebers has pointed out (Willer *et al.*, 2005):

> The abrupt change in risk by month suggests a threshold effect for both increased and decreased risk, something that is not easily explained. These observed changes may partly explain the increased risk of Multiple sclerosis in second generation Asian and Caribbean migrants to the United Kingdom – that is, moving to the United Kingdom does not change their genes but something in the climate may do so.

Ebers's studies and those of others such as Jim McLeod in Sydney show that in a population of individuals whose genetic background is very similar, there is a striking difference in risk depending on both when and where you are born.

There are several clues as to the cause. Half siblings who both have multiple sclerosis are far more likely to have the same mother than they are the same father. This parent of origin effect in itself does not differentiate between the maternal environment determining risk and what is known as genetic imprinting – in other words, if it is a gene, whom you inherit it from may matter – or of course it could be a combination of the two. However, the maternal parent of origin effect in this case suggests that it is the environment in the gestation or neonatal period that determines risk for this adult-onset disease.

An important study compared the risk of two or more siblings having multiple sclerosis with the risk of its occurrence in both twins in a non-identical pair. Because non-identical twins are no more similar genetically than any brother or sister born years apart, there should be no difference, but it turns out that non-identical twins do have a significantly higher risk that both will have multiple sclerosis than non-twin siblings. As twins are, by definition, born at the same time, give or take a few minutes or at most hours, the key difference here must be the prenatal and postnatal shared environment (Willer *et al.*, 2003).

However, pinpointing the actual environmental factor is tricky. The first clues to the interplay between the genome and the environmental factors during development came from observations in Scotland in the 1950s. Ebers relates (personal communication):

> Children who are born in the winter were at much greater risk from neural tube defects such as spina bifida and anencephaly and so forth. There is a strong association with lower socio-economic groups and a strong association with winter births. What was happening in Scotland in the 50s was that the poor could not afford fresh vegetables and at the end of the Scottish winter people tended to be deficient in folic acid. Now the evidence is very strong that you can prevent these birth defects by large amounts of folate. In this case we don't have anywhere near as strong effect that they had in neural tube effects but it does hint that the root to

preventing the disease may turn out to have a strong connection to maternal environment or perhaps very, very early in the neonatal period.

One strong signal linking vitamin D with multiple sclerosis is that in Japanese populations, who consume plenty of fish rich in vitamin D (90 per cent of the Japanese vitamin D requirement comes from fish, 3 per cent from eggs and 3 per cent from milk), the multiple sclerosis prevalence is 3 per 100,000 population, whereas in Scotland and Tasmania the prevalence is 250 per 100,000 population. This would tie in with the suggestion that the excess births in May imply that in Scotland and Tasmania the low-light levels of winter depressed vitamin D levels in pregnant women (Willer *et al.*, 2005).

Although the evidence that vitamin D is a protective environmental factor against multiple sclerosis is circumstantial, it is compelling. This theory can explain the striking geographic distribution of multiple sclerosis. It can also explain two peculiar geographic anomalies, one in Switzerland with high multiple sclerosis rates at low altitudes and low multiple sclerosis rates at high altitudes, and one in Norway with a high prevalence of multiple sclerosis inland and a lower prevalence along the coast. The intensity of ultraviolet radiation is greater at high altitudes, resulting in a greater rate of vitamin $D_3$ synthesis, thereby accounting for low rates of multiple sclerosis at higher altitudes. On the Norwegian coast, fish is consumed at high rates and fish oils are rich in vitamin $D_3$ (Hayes *et al.*, 1997).

The study of birth seasonality, despite its murky start in the wilder fringes of eugenics, is one of the tools that can be used to discover the effects of environmental factors. These studies on multiple sclerosis, using the gradients that show up in epidemiological data, may lead us to a better understanding of the interrelation between the internal and the external terrain. And they are important in their own right. George Ebers concludes that 'seasonal birth effect may be connected with environmental factors determining prevalence rates. These are powerful, seem to act at a broad population level and may hold the key to disease prevention' (Willer *et al.*, 2005).

For many millennia it has been believed that the season of birth has some relation to individual human destiny. Even though Shakespeare had Caesar proclaim that our faults lay within ourselves and not in our stars, nowadays far too many people still really believe that there is a correlation between the position of a planet such as Venus when we are born and our life chances. This kind of astrological idea is a remnant of a superstitious past that is fostered by commercial interests among a gullible public. And, no doubt, proponents of pseudoscience will leap on the realisation that birth month affects later life. This should not detract from the increasing recognition of the impact that changes in the maternal environment can have on later life of the offspring. Environmental conditions are intimately dependent on season, and as we are undoubtedly influenced by such conditions, the foetal or neonatal stages would probably be the most susceptible. It makes sense that the season of birth could be considered as a significant factor for our health and sickness. It is not just human health. Season of birth effects in domestic farm animals are so large and economically important that no breeder can afford to ignore them.

During the nine months from conception to delivery, the developing foetus is secure in the mother's womb. But it is not isolated from the world. The foetal physiology is enmeshed with that of the mother and can be challenged during the pregnancy with effects that may not be visible for years to come.

We readily accept that the circannual breeding rhythms of birds, sheep, hamsters and many other forms of life are synchronised by photoperiod in their natural surroundings, but the idea that modern humans may still be influenced, even if only subtly, by a similar process, with the consequences that our life-cycle health and well-being are dependent on our birth timing is intriguing and potentially important.

# 10

# DISEASE AND SEASONAL TIMING

Physicians should be aware of the 'Merry Christmas Coronary,' and 'Happy New Year Heart Attack'

ROBERT A. KLONER (KLONER, 2004)

The different seasons bring different illnesses. Half a century ago, summer, with its open-air swimming pools, was felt by many anxious parents to be an invitation to contract the dreaded poliomyelitis. It would be hard to convince people now, when they wilfully refuse MMR (measles, mumps and rubella) vaccine for their infants, of the heroic stature granted to Jonas Salk and his polio vaccine at a time when virtually every classroom contained at least one child with a leg in callipers. Table 10.1, compiled by Randy Nelson and colleagues, lists a wide range of seasonal diseases. The studies tend to be country-specific, and many of the findings have to be considered provisional.

We tend to get colds in winter and hay fever in spring. Food poisoning is more prevalent in summer, partly because bacteria grow faster in the hot, humid months. Malaria peaks just after the rainy season, when the mosquito population is rising. In nineteenth-century America, *Sommerkrankheit* ('summer illness') was the name given to cholera infantum, which was common during the hot summer months among hand-fed babies in most towns of the middle and southern states, as well as in many western areas. It was

| DISEASE | SEASON OF ONSET | REFERENCE |
|---|---|---|
| Malaria | Spring/summer | Hviid, 1998 |
| Legionnaires' disease | Summer | Fisman *et al.*, 2005 |
| Leishmaniasis | Winter/early spring | Andrade-Narvaez *et al.*, 2003 |
| Influenza | Winter/early spring | Zucs *et al.*, 2005 |
| Human retrovirus | Winter | Kapikian *et al.*, 1976 |
| Respiratory syncytial virus | Winter/early spring | Hall *et al.*, 1991 |
| Respiratory syncytial virus | Summer | Sakamoto *et al.*, 1995 |
| Coronaviruses | Winter/early spring | Cavallaro & Monto, 1970; Hambre & Beem, 1972; Hendley *et al.*, 1972 |
| Enteroviral infection | Summer | Glimaker *et al.*, 1992 |
| Tuberculosis | Winter | Pietinalho *et al.*, 1996 |
| Brucellosis | Spring/early summer | Dajani *et al.*, 1989 |
| Pneumonia | Winter/spring | Eskola *et al.*, 1992 |
| Mycosis | Winter/spring | Chariyalertsak *et al.*, 1996 |
| Coronary heart disease | | |
|   (a) Stroke | Winter | Douglas *et al.*, 1990 |
|   (b) Cerebral infarction | Spring/summer | Biller *et al.*, 1988 |
|   (c) Ischaemic attacks | Winter/spring | Wang *et al.*, 2002; Dunnigan *et al.*, 1970; Azevedo *et al.*, 1995 |
| Intracerebral haemorrhage | Winter/early spring | Azevedo *et al.*, 1995 |
| Transient ischaemic attacks | Summer | Sobel *et al.*, 1987 |
| Insulin-dependent diabetes mellitus | Autumn/winter | Blom *et al.*, 1989 |
| Rheumatoid arthritis | Autumn/winter | Rosenberg, 1988 |
| Childhood leukaemia | Winter | Karimi & Yarmohammadi, 2003 |
| Breast cancer | | |
|   (a) No. of cases diagnosed | Winter | Cohen *et al.*, 1983 |
|   (b) Initial detection | Spring/summer | Mason *et al.*, 1985 |
|   (c) Risk of death | No seasonal effect | Galea & Blamey, 1991 |
|   (d) Season of birth | Summer | Sankila *et al.*, 1993; Yuen *et al.*, 1994 |
| Lung cancer | Summer/autumn | Tang *et al.*, 1995 |
| Melanoma | Spring/summer | McWhirter & Dobson, 1995 |
| Urinary bladder carcinoma | Autumn/winter | Hostmark *et al.*, 1984 |

**Table 10.1** Seasonal incidence of multiple types of disease. (Modified from Nelson *et al.*, (2002).)

characterised by gastric pain, vomiting, purgation, fever and prostration. Death frequently occurred in three to five days.

Humans get ill all the time, but there are obvious seasonal highs and lows. As Hippocrates said in his *Aphorisms* some 400 years before the birth of Christ, 'All diseases occur at all seasons of the year, but certain of them are more apt to occur and be exacerbated at certain seasons.' It is not usually a simple relationship between the time of year and the weather conditions, but a complex interaction between the seasonal challenges to the body from outside and the seasonal cycling of our internal milieu. On his deathbed, Louis Pasteur, the founder of the germ theory of disease, allegedly said, 'the germ is nothing, the terrain is everything' (Delhoume, 1939). And it is an obvious point that two different people can be exposed to the same disease at the same time and one will become ill and the other not. Similarly, at different times of the year the same individual will react differently to the same challenge. Although every individual has a unique internal milieu, within each individual that milieu changes with the seasons and so does our susceptibility to disease.

Some illnesses, such as seasonal affective disorder (SAD), result from changes in our internal physiology that may be triggered by changes in the timing, intensity and amount of light; others result from external challenges from infectious agents such as the malaria and dengue fever carried by mosquito species; and for some it may be enabled by the weather – such as the warm, humid weather that foreshadows outbreaks of legionnaire's disease caused by a bacterium. Understanding these relationships may help prevent actual physical and mental illness and at the very least ameliorate the symptoms of any conditions that do occur.

The incidence of malaria peaks immediately after the rainy season. This has been known for millennia because malaria has been intensively studied since ancient times, if for no other reason than it has always killed large numbers of people. Nowadays it is a disease of the tropics, but the Greek physician Hippocrates described it in the fifth century BC. Malaria killed Alexander the Great in 323 BC. In the 1600s Jesuits brought the anti-malarial cinchona bark, later shown to contain quinine, to Europe, although Oliver Cromwell apparently refused the 'papist' remedy and died of malaria for his prejudices. The Nazis deliberately spread malarial infection as they retreated through Italy in the Second World War. Once common – it gets its name from the Italian *mala aria* ('bad air') – the World Health Organization only declared it officially eradicated from that country on 17 November, 1970.

Malaria kills an estimated one to two million people a year, mostly in sub-Saharan Africa, and the large majority of those are children. This disease of both poverty and climate is on the increase, and among the concerns about climate change is the fear that it will reappear in the higher latitudes. The *Plasmodium* protozoan that causes the disease in humans is transmitted by the bite of the female *Anopheles* mosquito (malaria is not directly contagious between humans except by transmission through the use of a shared hypodermic).

Transmission levels are influenced by environmental factors including temperature, rainfall and humidity. Malaria epidemics usually follow wet seasons. Mosquito mortality is high above 33°C, and below 18°C the development stages of the parasite inside the mosquito take longer than the mosquito's own survival, so transmission is disrupted.

People who live in areas where malaria is common can get the disease repeatedly and never fully recover, so there is a constant reservoir of *Plasmodium* available for transmission and thus to continue the cycle. The seasonal weather also determines human behaviours that may increase contact with *Anopheles* mosquitoes between dusk and dawn, when the *Anopheles* are most active. Hot and humid weather may encourage people to sleep outdoors or discourage them from using bed nets.

Mosquitoes are not only vectors for malaria. *Aedes aegypti*, the mosquito species famously recognised almost a century ago as the vector of yellow fever by the American army doctor Walter Reed, is also a carrier of dengue fever. Dengue is also spread by a close relative, *Aedes albopictus*. This species can tolerate cooler conditions, and these infected mosquitoes are spreading dengue from Mexico into the USA. The *Aedes* species also breed in still or stagnant water and their numbers increase towards the end of the rainy season.

The micro-organism that causes dengue belongs to a group known as flaviviruses. Many such as dengue, yellow fever and Japanese encephalitis are mosquito-borne, whereas others are transmitted by ticks. In most areas, transmission by arthropods and infection with flaviviruses only reaches epidemic proportions at the end of the wet season. However, as Jim Olson, a Texas Agricultural Experiment Station entomologist, reflects (Olson, 2005):

> We never run out of mosquitoes in Texas; we just change species with the season. We have some species that are active in the winter, in fall and summer, with the greatest number being the summertime species.

There are at least 86 mosquito species in Texas, classified into 13 genera. The seasonal specialisation is another example of dividing up ecological niches on the basis of time.

Trypanosomiasis is another seasonal killer. The protozoan *Trypanosoma*, which is transmitted by the tsetse fly (*Glossina* ssp.), severely constrains livestock and mixed crop–livestock farming in sub-Saharan Africa and causes sleeping sickness in people. Nearly nine million square kilometres of the land across 40 countries are infested with tsetse flies, and millions of cattle and people are exposed to the risk of disease. Although the disease was close to eradication by the mid-1960s, 40 years of war and civil unrest have severely curtailed the control programmes, and current estimates suggest that at least 100,000 people die each year.

The tsetse fly's life cycle is different from that of most biting flies that lay their eggs in a moist environment in which their larvae feed and develop. In the tsetse fly, a single fertilised egg is retained in the uterus during each pregnancy, where it is fed by the mother and develops until she deposits it as a late third-instar larva. The larva, which may weigh more than the mother, burrows into loose dry ground and within minutes is enclosed in a hard pupal case. The adult fly develops within this case and emerges, at least three weeks later, very hungry as it has not fed since its deposition as a larva.

Because of this life cycle, tsetse flies are not so much encouraged by wet conditions as inhibited by hot, dry conditions. Seasonal variations in numbers are generally smaller than in mosquitoes. Nevertheless, during the hot season in Zimbabwe for instance, when temperatures regularly approach 40°C, tsetse numbers decline by about 90 per cent (Torr & Hargrove, 1999). Seasonal changes in malaria and sleeping sickness, as well as trypanosomiasis in animals, play a large part in the annual migrations of nomadic peoples in sub-Saharan Africa. For example, the million or so Baggara spread over western Sudan and eastern Chad move with their animals to make the most of the available rainfall. But their migrations also need to be timed to avoid the beginning of the tsetse fly and mosquito season.

One of the major public health fears of climate change is that vector-borne diseases such as malaria will return to areas from which they were long ago eradicated and will perhaps move into regions where they have never been endemic before. Bill Bradshaw and Christina Holzapfel at the University of Oregon have worked for many years on *Wyeomyia smithii*, a mosquito that develops within the carnivorous leaves of pitcher plants. Their team have narrowed down the location of the genes that control the response to daylength to regions on three chromosomes. Two of the chromosomes also have an overlapping gene expression that tells the species to go into diapause, which they must do to survive. Bradshaw and Holzapfel believe that 'The map will allow researchers to narrow their focus to identify specific genes that control the seasonal development of animals' (Mathias *et al.*, 2007). This offers some hope that it may be possible to predict which animals may survive in changing climates and identify which disease-carrying vectors may move northwards.

Heart disease is another of the killer illnesses with a marked seasonality; however, its causes are not usually external pathogens but in the main diet, smoking and other lifestyle factors, and stressors allied to genetic predispositions. Whether the location is London, New York or Tokyo, there are 50 per cent more heart attacks during winter than at any other time of year, which is perhaps not unconnected with higher cholesterol levels in winter or the fact that winter blood pressure in northern latitudes, as in Britain, is about 5 mmHg higher than during the rest of the year.

Winter heart attacks have a higher fatality rate. In England and Wales, the winter period accounts for an additional 20,000 heart-attack deaths each year (Pell & Cobbe, 1999). One cause could be that in the cold, the blood vessels constrict to help conserve body heat. A consequence of narrowed vessels is higher blood pressure, which puts an additional strain on the heart. However, it is not the snow that triggers an attack: it is the physical stress of shovelling it away that does the damage. Or it could be influenza or other infections. These are more common in the winter months – even in tropical climates. In one study, researchers found that the risk of a heart attack was temporarily tripled in the ten days after an acute respiratory tract infection (Meier *et al.*, 1998). And in a study of deaths in Los Angeles County, the greatest number of heart attacks occurred around 1 January. Because the weather in this part of the world is mild all year round, it is likely that it is the emotional stresses of the holidays, coupled with people's tendency to overindulge in food, alcohol and sex during that time that may be factors. A high-fat holiday meal can interfere with relaxation of the arteries and may also activate the clotting system, which can spell trouble for people with coronary artery disease. Also, excess alcohol intake can increase blood pressure and contribute to heart rhythm abnormalities (Kloner, 2004). Add in the exertions of sexual activity and it is a potent brew, although it could just be that there are more delays at this time of year in seeking treatment and so the heart attack victims who arrive at the Accident and Emergency departments are in a worse state than they might otherwise have been (Phillips *et al.*, 2004).

In their review of the seasonal factors that could affect heart attacks, Jill Pell and Stuart Cobbe of Glasgow University note that, apart from those factors already mentioned, researchers have looked at the following: ultraviolet radiation and vitamin production; haemoglobin levels; glucose and insulin levels, which are lower in summer than winter; peptic ulceration caused by the bacterium *Helicobacter pylori*, which also has a winter peak; age; sex; and a raft of others.

Untangling them all is a major effort, and as Pell and Cobbe point out (Pell & Cobbe, 1999):

It is unclear whether the excess deaths which occur in winter reflect avoidable deaths or merely slightly premature deaths which would have occurred anyway within a short period of time. Identifying and rectifying those factors associated with seasonal variations will only impact on overall mortality if the former is true.

Malaria, sleeping sickness and heart disease can kill. Colds by themselves usually just make life unpleasant. But whereas it is easy to see from biological life cycles why malaria and sleeping sickness and other vector transmitted diseases are seasonal, it is not at all clear at first sight why colds generally occur in winter. The cocktail of 200 or more viruses that are responsible for variants on the common cold, including rhinoviruses and coronaviruses, are around in the air more or less all of the time. The textbook answer is that there are more colds in cold weather because we tend to crowd indoors in poorly ventilated rooms and that makes transmission easier. But common colds are relatively hard to catch. Spreading the infection requires close and prolonged contact with other people. The viruses replicate in the cells lining the nose and are coughed or sneezed out in droplets of mucus. Secretions from the nose can also be spread on our fingers and contaminate others.

The crowding theory for the common cold has been around for more than 100 years, but our cities are just as crowded in summer as in winter. In earlier times, humans did live in close proximity with their domesticated animals in winter, a practice that still occurs in poor, rural areas, particularly in the developing world, and this may well account for the development of new influenza strains originating in these places in their winter.

According to the Common Cold Laboratory at Cardiff University, a new theory (Eccles, 2002) to explain the seasonality of colds puts forward the idea that

> Our noses are colder in winter than summer and that cooling of the nose lowers resistance to infection. If the weather is freezing outside we wrap up in winter clothes but we still leave our nose exposed to the freezing air. Every time we breathe in we cool the nasal lining and weaken our local defences against infection. If this theory is correct then covering our nose with a scarf in cold weather could help prevent colds.

Another possibility is that cold air and low relative humidity are the cause. The virus is most stable at a relative humidity of between 20 and 40 per cent. The dry air leads to smaller water droplets on which viruses are carried, enabling them to remain airborne for long periods and so enhancing the likelihood of spreading the infection. Further, cilia in the respiratory system work more slowly in the cold, enabling the virus to spread in the respiratory tract and to disperse in a sneeze or a cough (Lowen *et al.*, 2007).

There is also considerable evidence that the stress of everyday life suppresses the general resistance to infection and can influence susceptibility. When common cold viruses have been administered to the nose of healthy volunteers, there is a link between a recent history of psychological stress and susceptibility to infection. Psychological stress affects the immune system, and the most likely cause may be the increase in the corticosteroid hormones associated with stress, as the corticosteroid hormones suppress the immune system (Cohen, 1995).

Rotavirus, which is responsible for most childhood diarrhoea, is highly seasonal. Each year the disease kills more than 600,000 children, mainly in the developing world, and virtually every child in the world contracts a rotavirus infection at least once before reaching five years of age. Just ten virus particles can start trouble in a young child. In the 1980s, researchers in the USA realised that infection follows a distinctly seasonal pattern in that country. Rotavirus gastroenteritis appears in the southwestern USA and then slowly migrates to northeastern cities such as Boston and Washington DC, striking young children during December to March.

Although the weather conditions are intimately linked with many diseases, there are ostensibly seasonal illnesses that are hard to pin down to regular changes in atmospheric conditions. In sub-Saharan Africa, meningococcal epidemics closely follow the season of dry winds and end with the onset of the rains. One explanation was that the drying of mucosal surfaces increases the probability of bacterial spreading and that the rains moisten the mucosa or decrease the spread of the organism by dust. However, in Oregon in the USA and other areas, meningococcal disease peaks during the rainy season. Similarly, a significant correlation between the onset of the invasive pneumococcal disease season and a decrease in mean daily temperatures below 24°C in Houston was not confirmed in seven other areas with more widely varying weather patterns. Respiratory syncytial virus (RSV) epidemics occur in the colder months of winter and spring in the USA, and commonly cause bronchitis and mild infections of the upper respiratory tract that can result in severe, and even fatal, lower tract problems, especially among infants. Paradoxically, RSV epidemics are significantly correlated with the hotter months in Singapore and Hong Kong (Dowell, 2001).

The relationship between illness and seasonal factors may turn out to be very complex, but the potential benefits of understanding the connection between our internal physiology and susceptibility to external agents of disease, whatever their source, is exciting researchers. The question is changing from why seasonal infectious disease outbreaks occur when they do to why they do not occur when they do not. If the annual troughs of summer influenza or winter hay fever are due to increased host resistance, and we can get a handle on the mechanisms, then new therapeutic pathways may open up because the swing between peak and trough can be very large. Frank Bronson summarises this key point when he states, 'Individuals have evolved mechanisms to bolster immune function in order to counteract seasonally recurrent stressors that may otherwise compromise immune function' (Nelson *et al.*, 2002).

Scott Dowell, of the Centres for Disease Control and Prevention in Atlanta, Georgia, has noted (Dowell, 2001):

> Seasonal cycles of infectious diseases have been variously attributed to changes in atmospheric conditions, the prevalence or virulence of the pathogen, or the behaviour of the host. Some observations about seasonality are difficult to reconcile with these explanations. These include the simultaneous appearance of outbreaks across widespread geographic regions of the same latitude; the detection of pathogens in the off-season without epidemic spread; and the consistency of seasonal changes, despite wide variations in weather and human behaviour.

By way of illustration, Dowell points out that influenza outbreaks across North America and Europe tend to be simultaneous, happening too quickly simply for person-to-person transmission from a single initial source. Regular changes in susceptibility may be a better bet.

There is a general hypothesis that explains our susceptibility to colds as the result of the need to 'steal' energy from other systems, including our immunological defences, to maintain our body temperature and other metabolic processes in cold weather. The idea is that the human immune system is tuned to the predictable seasonal change in external stressors and is enhanced in anticipation of colder temperatures. But if the challenge is too severe, or if the enhancement of the system is insufficient, then the compromised immune system cannot cope with the challenge of the virus or other infectious agent.

The shift in focus suggests that when challenged by regular seasonal variance, there is an evolutionary advantage in modulating the immune system so that it is best placed to counteract such stressors, given the energetic constraints in so doing. The energy cost–benefit rider is important. Although the ideal strategy would be to have an immune system operating at maximal level at all times, the energy cost in doing so is too high and would be at the expense of other functions such as growth and reproduction.

Virtually anything that challenges an organism's homeostatic balance is a stressor. Stressors cause stress, and this can affect the immune system in humans. We think of pneumonia as a disease of old age. To a generation of doctors before the Second World War it was known as the 'old person's friend' because in the absence of antibiotics it carried them off quickly. But it has also been a problem for the military as fit young men have gone down with it. In two weeks in March 2002, there were 31 cases admitted to just one military hospital in southern India. The Army doctors attributed the outbreak to stress caused by overcrowding in the barracks depressing the recruits' immune system and so heightening their susceptibility to the disease (Banerjee *et al.*, 2005).

Other studies on military personnel support the idea that stress, for instance exercise, compromises immune function. Cases of pneumonia among US Ranger Corps prompted a study of their training regimes. In their two months of initial intense training, recruits could lose between 12 and 16 per cent of their body mass. This weight loss affected their T-lymphocyte response, which was restored within nine days when the recruits boosted their calorie intake by 15 per cent (Kramer *et al.*, 1997).

Intriguingly, it is not necessarily that the stressors themselves should interact directly with the immune system. The very perception of stress can do it. Robert Sapolsky, writing in his inimitable manner, describes it thus (Sapolsky, 2003):

> Primates have it tough, however. More so than in other species, the primate stress response can be set in motion not only by a concrete event but by mere anticipation. When this assessment is accurate ('This is a dark, abandoned street, so I

should prepare to run'), an anticipatory stress response can be highly adaptive. But when primates, human or otherwise, chronically and erroneously believe that a homeostatic challenge is about to come, they have entered the realm of neurosis, anxiety and paranoia.

Sapolsky has described one experiment in which two monkeys were under-fed, but one monkey received non-nutritive food while the other received nothing. Despite the fact that the food was non-nutritive and both monkeys were equally underfed, the one that received the placebo did not display the elevated glucocorticoid levels that are evidence of stress, whereas the other did.

A better understanding of the mechanisms underlying seasonality and weather in the occurrence of infectious diseases is of more than academic interest. David Fisman of the Drexel University School of Public Health in Philadelphia has pointed out, appropriately enough given the location (Fisman *et al.*, 2005):

> Recognition of weather and seasonal patterns in occurrence of diseases such as legionnaires may enhance the likelihood that clinicians recognize this under-di-agnosed disease, improve the accuracy of 'syndromic' surveillance systems cur-rently in use for identification of bioterrorist attacks, and inform public-health policy in the context of global climate change.

Although many modern humans live in a world buffered from such stressors as food shortages or cold, there are still seasonal changes in popu-lation-wide disease and death rates and a concomitant cycling in the im-mune system. Stressful environmental conditions affecting animals – such as reduced food availability, low ambient temperatures, overcrowding, lack of shelter, or increased predator pressure – recur seasonally, as do seasonal fluctuations in immune function among individuals. However, to under-stand the interaction fully needs some background on the immune system and its relationship to the neuronal and endocrine systems as well as an understanding of the energy balances that have to be maintained by every living creature.

The study of the human immune system has had a chequered research history. In 430 BC, Thucydides recorded that while the plague was raging in Athens, the sick and dying would have received no attention had it not been for those individuals who had already contracted the disease and recovered and recognised what we would later call their 'immune' status. Variolation, a form of inoculation, was known to the Turks in the Middle Ages, and in 1798 Edward Jenner injected a young boy with cowpox and showed that he acquired immunity to the deadly smallpox.

However, despite the long history, Louis Pasteur can probably be considered the father of immunology. On 6 July 1885, the parents of nine-year-old Joseph Meister brought the boy to Paris. Their son had been severely bitten by a rabid dog two days earlier. Pasteur injected attenuated rabies virus into the young lad. Joseph Meister survived, and this was the first known case of an individual being bitten and surviving rabies.

Notwithstanding the romance of Meister's story and the recognition of the importance of vaccines, the immune system, with its confusing terminology, remains difficult to understand. But it is of critical importance to our well-being. A key question in medical research for the past century has been immunological: How does the body defend itself against microbial pathogens?

The immune system can be thought of as a string of forts stationed around the body protecting it against hostile incursions. Linking the forts together is a network known as the lymphatic system, and the forts are the lymph nodes. The lymph system interconnects with the circulatory system and complements the blood supply. Lymph itself is a clear, watery fluid derived from body tissues. It contains white blood cells and circulates throughout the lymphatic system filling tissue spaces and distributing the nutrients, water and oxygen that those individual cells need. (Blood transfers these materials to the lymph through the capillary walls, and lymph carries it to the cells. Even though every cell in the body is no more than five cells away from a capillary, without the lymph system each cell would have to have its own blood capillary supply.)

Playing the part of the grizzled scouts patrolling the system are cells known as macrophages. They are on the lookout for viruses, bacteria, fungi, parasites – anything that causes harm. Helping out are neutrophils, which we can think of as junior scouts. Neutrophils are by far the most common form of white blood cells; the bone marrow produces trillions of them every day and releases them into the bloodstream. Once a neutrophil finds a foreign particle or a bacterium it will engulf it, releasing enzymes, hydrogen peroxide and other chemicals from its granules to kill it. In a site of serious infection (where lots of bacteria have reproduced in the area), pus will form. Pus is simply dead neutrophils and other cellular debris. Macrophages develop from cells in bone marrow that are then released to float in the bloodstream, enter tissue and turn into macrophages. They are a bit like amoebae in the way they make contact with the invader cell and counter-attack by dissolving, weakening, engulfing, digesting and eliminating it.

If there are too many invading cells, or if they are of a novel type, a macrophage can tag an invader and recruit more specialised types of immune cells to enter the battle. It sends a chemical signal known as interleukin-1 (think of John Wayne despatching a young trooper), which summons help in the form of B cells, which are made in the bone marrow, and T cells, which are also made in bone marrow but mature in the thymus.

This is where it starts to get really complicated. T helper cells can send a signal in the form of interleukin-2, which triggers T-cell proliferation; this in turn causes cytotoxic killer cells to turn up and destroy the infectious agent. This route is termed cell-mediated immunity.

But there is another route known as antibody-mediated immunity. The T-helper cells send a B-cell growth factor signal instead, which causes differentiation and proliferation of B cells. These make and release specific antibodies that bind to the infectious agent and target it for destruction by a group of proteins known as complement.

In truth, it is far more complex than this, with many more components such as leukocytes, lymphocytes, monocytes, granulocytes, plasma cells, killer T cells, suppressor T cells, natural killer cells, eosinophils, basophils and phagocytes. And those are just the cells. Apart from cell growth factor, there are interferons, tumour necrosis factor and a bunch of hormones including the interleukins, known as lymphokines, monokines and chemokines.

When macrophages and other cells are out on patrol they engage in skirmishes; the whole party meets up in the lymph nodes, where the full-pitched battles are fought between the defenders of the immune system and the invaders.

This wondrously effective but complex system hooks up with the neural system via the hypothalamus, which explains why how we feel affects how well our immune system is working.

The point about all this complexity is its dynamism. Cells and chemical messengers are continually being made and destroyed to maintain an effective immune system, and all this activity takes up energy. Partitioning energy to different and often conflicting activities – growth, reproduction, thermoregulation and thermogenesis, healing, locomotion, hibernation, moulting, vision and everything else that may be necessary to maintain existence – is the ultimate survival problem. Although it would be ideal to maintain the immune system in a high state of readiness all year round in animals including humans, the energy costs of doing so are very high.

Each minute of every 24 hours, our bodies are allocating resources to different functions. For instance, the digestive process itself accounts for 10–20 per cent of energy needs. And special circumstances make special demands. During pregnancy, a woman uses about an extra 50,000 calories. When nursing, she needs 500 or more calories a day just to produce an adequate supply of milk.

A classic study into energy allocation in humans was conducted towards the end of the Second World War as the Allies encountered starved, emaciated civilians, many of whom had survived by subsisting on bread, potatoes, and little else. Relatively little was known scientifically about human starvation or how to deal with re-feeding people who had undergone this extreme degree of deprivation.

Ancel Keys, a physiology professor at the University of Minnesota, designed and led what became known as the Minnesota Starvation Experiment. Thirty-six young men, all of them conscientious objectors, volunteered for the study. They spent the first three months on a normal diet. For the next six months their calorie intake was cut in half, and for the last three months they went through a re-feeding protocol.

On average, the fit young men lost 25 per cent of their body weight during the 'starvation' period. As they weighed about 70–80 kg at the start, by the end of the six months they were emaciated. The men reported decreased tolerance to cold. They experienced dizziness, extreme tiredness, muscle soreness, hair loss, reduced coordination, and ringing in their ears. Several of the men said that almost immediately after semi-starvation began, all interest in women and dating was lost: 'I can tell you, the sex drive disappeared. There was none' (Kalm & Semba, 2005). Another put it more graphically, stating that he had 'no more sexual feeling than a sick oyster' (Garner, 1997).

Among the physical changes were decreases in body temperature, heart rate and respiration, as well as in basal metabolic rate – the amount of energy (in calories) that the body requires at rest (that is, in the absence of physical activity) to carry out normal physiological processes. It accounts for about two-thirds of the body's total energy needs, with the remainder being used during physical activity. At the end of semi-starvation, the men's basal metabolic rates had dropped by about 40 per cent from normal levels.

The subjects of the study had adapted to the reduced energy input by switching resources so that they could manage on fewer calories. But in doing so they had compromised many physiological functions that 'in the wild' were essential to survival.

Ancel Keys did not have today's technology and so could not track changes in the immune system during his studies, but in a series of studies beginning in the mid-1970s Alain Reinberg found that components of the immune system in humans show an endogenous seasonal variation and that among healthy subjects there were significantly more circulating lymphocytes in the winter months (Reinberg et al., 1977).

There are more of the lymphocytes that are involved in antibody-mediated immunity, the B cell and helper T-cell route, in winter, whereas the converse seems to be the case with lymphocytes such as cytotoxic T-cells, which are involved in cell-mediated immunity.

There may be seasonal trade-offs between antibody-mediated and cell-mediated immunity. Nelson and his colleagues have pointed out (Nelson et al., 2002):

Antibody-mediated immunity is essential for overcoming bacterial infections. Bacterial infections are most prevalent in the winter. Thus, seasonal variation in antibody-mediated immunity may have evolved as a co-adaptation to seasonal changes in bacterial pathogens. Conversely, cell-mediated immunity is higher in the summer and is critical for killing viral pathogens.

The jury is still out as to the extent of seasonal tuning of the human immune system. It may be that the seasonality of the pathogens invokes the immune system rather than that the immune system is anticipatory. It may be a mixture of anticipation and response. As yet, we do not have definitive evidence that the number of cells or the general state of competence of the immune system changes in advance of periods of peak seasonal challenge. The problem is that most humans now live in artificial environments, mainly indoors, and we are buffered from extremes. So in humans we have to look for relatively small and subtle changes in fairly small numbers of subjects, and this makes it difficult to come up with clear-cut results.

There is an intuitive appeal in the notion that the immune system rises and falls with a regularity that mirrors the rise and fall of external challenges, and this seasonal variation is the basis of many systems of folk medicine. Chinese medicine focuses on balancing what it calls positive and negative forces within an individual, allowing the patient's own immune system to keep the body healthy. The idea is that there is a cyclical flow of a person's Qi (pronounced chi) or inner energy corresponding to the seasons. So winter is regarded as a time of withdrawal from the world to reflect, rejuvenate and rest. When the season shifts into spring, energy bursts forth, and as daylight lengthens, the inwardly contained energy begins to flow instead of being stored. Traditional Indian Ayurvedic medicine has a similar focus on the individual and the idea of balance and harmony. As in Chinese medicine, the change from one season to another requires shifting one's diet to restore balance.

A seasonal fluctuation of the competence of the immune system may be of considerable importance in the light of coming environmental threats. Animals and plants are already increasing their geographical range as the temperature rises. Mosquitoes may well return with a vengeance to areas that have been free of their associated diseases for years. We also face an increasingly overcrowded world with pressure on water supply and sanitation, which may lead to increases in the intensity of seasonal patterns of infection. We are not yet at a stage at which broad swathes of the current temperate zones will have to contend with seasonal outbreaks of malaria, sleeping sickness, dengue fever, cholera and other dread diseases, but it is within the realms of possibility.

It is not as though seasonal change influences only the immune system. Our moods, performance, sleep patterns, thermoregulation capability, thyroid function, cortisol levels and almost everything else one cares to mention show some seasonal variation. Anna Wirz-Justice has provided a full list, and from it she concludes (Lacoste & Wirz-Justice, 1989):

> Whether the measures are of behaviour, psychology, hormones or neurochemistry our review emphasizes their relatedness with respect to season. Yet oscillatory behaviour is not pathological, but constitutes a functional advantage. For seasonal rhythms, as for circadian rhythms, the ability to predict repetitive events with appropriate physiological responses is of evolutionary significance.

# 11

# SEASONAL AFFECTIVE DISORDER

'of constitutions some are well or ill-adapted to summer, others are well or ill-adapted to winter'

HIPPOCRATES (HIPPOCRATES, 400 BC)

Some of us love the long, gloomy days of winter in the higher latitudes, but most of us brighten up in the sunlight. Get off the plane in the middle of a bright Australian summer day and no matter how jet-lagged you are your spirits rise, though it does not prevent about one in four Australians from suffering clinical depression at some point in their lives. It has been appreciated for millennia that light in itself is therapeutic for what was called melancholia, and there is a dose–response effect in that the more light you get, the better you feel (Cajochen, 2007).

In 4000 BC, the Babylonian king Hammurabi ordered his priests to use sunlight in the treatment of illnesses (Koorengevel, 2001). The ancient Egyptians, Babylonians and Assyrians had their sun-gardens. Sunlight was promulgated by Hippocrates and succeeding Greek and Roman physicians. Working as a physician in Rome in the first century AD, Celsus advised his patients to live in rooms full of light. At about the same time, Aretaeus of Cappadocia, who can lay some claim to having been the first psychiatrist, was advocating light therapy, writing that 'lethargics are to be laid in the light and exposed to the rays of the sun (for the disease is gloom)' (Eagles,

2004). The Germanic tribes employed sunbathing, and the epic poem Edda tells us how they carried their sick in the springtime to the sunny slopes to expose them to the sunshine. Some tribes placed their feverish children on the tops of houses in the sunlight, so that they might recover more rapidly (Fielder, 2001). With the rise of the Christian Church and its decidedly hostile stance to 'sun-bathing' as a pagan practice, its use fell into disfavour. However, in the early nineteenth century, the French physician J. F. Cauvin wrote his PhD thesis at the University of Paris on the benefits of sunlight, and he recommended the prescribing of sunlight to what he described as the sad and the weak. He also stated that light was a curative agent for scrofula, rickets, scurvy, rheumatism, paralysis, dropsy, swellings and muscle weakness (Cauvin, 1815).

Sunlight has a history as a powerful medicine, although few would advise going as far as the authors of *Sunrays and Health*, who in 1929 advised modern women that 'Instead of hiding under parasols to protect pink and white skins for evening wear, they lie by the hour on the unshaded beach developing all tones from pale yellow to saddle colour' (Millar & Free, 2004).

There are claims that the first description of what we now call seasonal affective disorder (SAD) is found in Jordane's *Gotica*. Written in about AD 550, this account of the history and geography of the Goths includes a comment on the Adogit people, who lived in 'Scandza' or Scandinavia. Jordanes recounted (Jordanes, AD 551) that the Adogit were said

> to have continual light in midsummer for forty days and nights, and likewise no clear light in the winter season for the same number of days and nights. By reason of this alternation of sorrow and joy they are like no other race in their sufferings and blessings.

In his account of the development of attitudes to seasonal depression, Tom Wehr of the US National Institute of Mental Health gives due prominence to Philippe Pinel, known as the father of modern psychiatry, who became director first of the Bicêtre hospital during the French Revolution and soon thereafter of the 7,000-patient Salpêtrière institution (Wehr, 1989). Pinel described three cases in his *Treatise on Insanity* (Pinel, 1806), in which the paroxysms returned upon the approach of winter, i.e. when the cold weather of December and January set in; and their remission and exacerbation corresponded with the changes of temperature of the atmosphere from mildness to severe cold.

Pinel stressed the importance of the changes of the seasons and the weather on the condition of psychiatric patients and stated that periodic insanity was the most common form of psychiatric disease.

One of Pinel's students, Jean-Étienne Esquirol, who became his successor, continued his teacher's investigation into seasonal effects on mood. In 1825 he began treating M, a 42-year-old man, at the close of winter, and Esquirol recorded the man's tribulations (Esquirol, 1845):

> Three years since I experienced a trifling vexation. It was at the beginning of autumn and I became sad, gloomy and susceptible. By degrees I neglected my business and deserted my house to avoid uneasiness. I felt feeble ... I became irritable .... I suffered also from insomnia and inappetence .... At length I fell into profound apathy, incapable of every thing, except drinking and grieving. At the approach of spring I felt my affections revive. I recovered all my intellectual activity and all my ardour for business. I was very well all the ensuing summer, but from the commencement of the damp and cold weather of autumn, there was return of sadness, uneasiness ....

Esquirol's account is a fair description of what nowadays is known as winter SAD. Furthermore, he also noted that SAD might occur in a subsyndromal form, what we now call S-SAD, or more popularly 'winter blues'. For those with the means, Esquirol prescribed a stay in Italy from October to May, which no doubt did wonders for the Italian tourist industry.

The winter depression probably affects anyone, but claims are made that it is particularly severe on creative types. Writers seem to have had a bad time of winter. Charles Kingsley, best known as the author of *The Water Babies*, hated it (Kingsley, 2004):

> Every winter, when the great sun has turned his face away
> The earth goes down into a vale of grief,

And fasts and weeps and shrouds herself in sables,
Leaving her wedding garlands to decay
Then leaps in spring to his returning kisses.

Nor did the great nineteenth-century American lyric poet Emily Dickinson betray any love for the season when she wrote (Dickinson, 1999):

There's a certain Slant of light,
Winter Afternoons –
That oppresses, like the Heft
Of Cathedral Tunes –

Victorian polar explorers complained about the languor brought on by winter darkness. One of the most controversial of this redoubtable breed was Frederick Cook, an American physician, who was described by Roald Amundsen, no less, as the most remarkable man he had ever met. Cook is regarded by many as being the first person to have reached the geographical North Pole. He spent several years on the northwest coast of Greenland, and he described a syndrome among his shipmates and the indigenous population characterised by depressed mood, fatigue, and loss of energy and sexual desire. He wrote:

The light of the Arctic summer is ... an efficient tonic to the mind and body; but before the night begins the stimulation is replaced by a progressive depression. The darkness, cold and isolation then drive the mental faculties on to melancholy.

As if his time in the high northern latitudes was just a taster of things to come, Cook then went to the Antarctic and in 1898 he apparently used bright lights to increase the well-being of the ship's crew while trapped in the ice (Rosenthal & Blehar, 1989).

However, if there is to be an acknowledgment to the first 'scientific' description of what became known as SAD, it could reasonably go to the German psychiatrist Wilhelm Griesinger. Writing in 1855, Griesinger described 'cases where regularly at one particular season – for example in winter – a profound melancholia has supervened, which in spring passes into mania, which again in autumn gradually gives way to melancholia' (Eagles, 2004).

In the twentieth century, another German psychiatrist, Helmut Marx, described episodes of recurrent winter depression among soldiers fighting in the Second World War in the cold, dark reaches of northern Scandinavia. He treated them successfully with sunray lamps and open-air light baths (Marx, 1946). Marx hypothesised (rather remarkably for the times) that these symptoms were either due to a pituitary insufficiency or *simply related to the influence of the altered 24-hour light/dark cycle on the hypothalamus* (our italics).

The modern story of winter SAD starts with Herb Kern, an American research engineer. Kern had kept copious notes of his regular seasonal emotional cycles for 15 years and he realised that he typically suffered depressions associated with winter and episodes of hypomania in the spring.

He had developed his own idea that as his mood changed with the lengthening and shortening of the days, his episodes of bipolar affective disorder may have had something to do with environmental light. He joined the American Society of Photobiology, attended meetings and he got to know investigators in the subject. When he learned of the work on light and melatonin of Al Lewy, Tom Wehr and other members of the team at the National Institute of Mental Health (NIMH) facility in Bethesda, Maryland, he contacted the NIMH group and told them about his illness and his ideas about light.

Lewy and Wehr had already shown that in humans bright light suppresses the night-time production of melatonin (Lewy *et al.*, 1980). This was an important finding, because until then it was thought that humans, unlike other mammals, responded to social cues rather than light to regulate their timing behaviours.

In the winter of 1980–81, the 63-year-old Kern was admitted for treatment. Lewy reasoned that by attempting to mimic summertime through artificially lengthening the day and so suppressing the melatonin levels, it might be possible to alleviate the depressive symptoms (Lewy *et al.*, 1982). He proposed that Kern should receive six hours a day of bright light, three before dawn and three after dusk. What Lewy and his team were doing was

| TYPICAL RANGE (LUX) | SITUATION |
|---|---|
| 100,000 | Bright sunny day |
| 10,000 | Cloudy day |
| 1,000–2,000 | Watch repairman's bench |
| 100–500 | Typical office setting |
| 200–1,000 | Night sports field |
| 10 | Twilight |
| 1 | Deep twilight |
| 0.1 | Full moon |
| 0.01 | Quarter moon |
| 0.001 | Moonless clear night |
| 0.0001 | Moonless overcast night |

**Table 11.1** Amount of light under different situations measured in lux (a measure of light that approximates 'brightness' as observed by a human subject). Source: the Lighting Research Center.

novel in the sense that they thought that it was the combination of the intensity of the light, the duration and timing of the dose that mattered.

The NIMH researchers sat Herb Kern in front of a home-made light box housing full-spectrum fluorescent lamps that delivered a light intensity that equated to that which someone would receive 'standing at a window on a spring day in the north-eastern United States' (Rosenthal, 2006). This was equivalent to about 2,000 lux. (One lux is about the illuminance provided at one metre by an ordinary wax candle.)

In general, we receive surprisingly little light because we seldom spend much time outdoors nowadays. Table 11.1 shows that in an office we receive some 20–40 times less light than we do in daylight, and about 200 times less than if we are on a sunny beach in mid-summer. During the winter months, people working in an interior office away from a window are potentially in biological darkness all day.

Within three days, Kern began to feel better. According to Norman Rosenthal, a young South African doctor working on the NIMH team, the change was dramatic and unmistakable. After the first successful treatment, nine patients were treated the next year with bright light and dim light exposure at the same times. Lengthening the day with bright light was an effective antidepressant; dim light had no effect.

By the spring of 1984, Rosenthal and his colleagues had identified and treated a total of 29 patients with what they called seasonal affective disorder, a syndrome characterised by recurrent depressions that occur annually at the same time. Individuals suffering from SAD tend to tire easily, crave carbohydrates, gain weight, experience increased anxiety or sadness, and show a marked decrease in energy. With protracted daylight in the spring, patients emerge from their depression and sometimes even display modest manic symptoms (hypomania).

Over the next 20 years the NIMH team worked hard to persuade the medical world that SAD was a genuine condition and that many of those who suffered from it could be treated with bright light therapy. In some respects they succeeded, and light boxes for SAD are on sale in pharmacies, department stores and other outlets. But in other ways they failed. In the USA, most insurers do not offer reimbursement for bright light treatment for SAD, and most medical courses and residency programmes do not provide clinical training in phototherapy. As Anna Wirz-Justice, a leading researcher based in Switzerland, has pointed out (Wirz-Justice, 1998):

the biological psychiatry establishment has regarded light therapy with a certain disdain and relegated it to the edge of the paradigm – not molecular enough, a bit too Californian – alternative, a bit too media overexposed, merely a placebo response by mildly neurotic middle-aged women who don't like nasty drugs.

The NIMH team has been disbanded, and Rosenthal remarked in 2005 (Moran, 2005):

Research on light therapy today has come to a virtual standstill because of a lack of funding, and thus the therapy has been marginalized. Federal funding that had once been generous has dried up, and since equipment used for light therapy does not hold the promise of yielding huge profits for manufacturers, other sources of funding have not materialized.

I know of no published data that speak to the frequency with which clinicians prescribe light therapy, but my impression, based on many years of working with SAD patients, is that clinicians often just don't think about it as a legitimate therapy or else fail to prescribe light for other reasons.

Is light therapy simply too hippy-dippy for the conservative, medical profession or are there other and more substantive reasons as to why a seemingly safe and effective treatment has been unable to find a place among mainstream practitioners?

For two decades there has been a debate as to whether SAD really is an illness, and whether people who suffer from it are seasonal depressives or whether they are recurrent depressives who happen to usually feel worse during winter. After all, our moods vary according to events in the world around us. We are happy when we achieve something or saddened when we consider we have failed. We use the word 'depressed' somewhat loosely, but the clinical depressions that are seen by doctors differ from the low mood brought on by everyday setbacks. Depression itself has a long history. King Saul is described in the Old Testament as experiencing depression and committing suicide because of it. What we call clinical depression is a common illness with a lifetime risk in the developed world of between 10 and 20 per cent; the incidence in women is two to three times higher than it is in men, though this does raise some questions about recognition of the condition and presentation to a physician. With so many people suffering from depression, some will show a winter pattern over two, three and even four succeeding years even if the illness is randomly distributed over the year. If we are talking about less severe depressions, then the incidence will be somewhat higher and the random nature of its timing will be even more likely to throw up what seems to be seasonal occurrences.

There are some who dismiss the whole notion of SAD. They argue that those of us living in higher latitudes feel a bit low in winter, what with the rain and colds and 'flu, and even if it does exist, why should it be viewed as an 'illness', and if it is an illness, why should it be seen as any different from standard depression? Others argue that it is a real condition with symptoms that can be misread as several other conditions, such as glandular fever (mononucleosis), hypoglycaemia or hypothyroidism.

Perhaps nowhere is more attuned to seasonal change and its effects on mood than Tromsø, a Norwegian city just north of the Arctic Circle. The sun sinks below the horizon from November 21 to January 21, and there is constant daylight from May 21 to July 21. Humans have been living in the area for a long time, and there is evidence of Saami culture 2,000 years ago. For all that time, humans have learned to cope with extreme changes in seasonal conditions. They have used social and cultural means to enable them to live through the dark winter nights and there is some resistance to what is seen as a modern attempt to medicalise what is considered to be the natural order. The people living in northern Norway consider themselves adapted to the seasonal changes and look askance at their southern countrymen, who seem to be more akin to mainland Europeans in their approach to the dark days of winter and the attendant discomforts. However, when the population of Tromsø fill in questionnaires about how they feel, up to 25 per cent are self-rated as suffering from SAD!

It may help that children in northern Norway are dosed up every day with cod liver oil, which is rich in omega-3 fatty acids, and this becomes a lifetime habit in many cases. Although still controversial, there is strong evidence from epidemiological studies that there is a low incidence of depression in cultures in which large amounts of fish containing these oils are eaten. This potential confounding effect illustrates how difficult it can be to pin down the causes of illness, particularly in mental health.

There are some among what might be called the 'green movement' who challenge the very existence of a condition such as SAD along the following lines (Smith, 2005):

> So what if we slow down as winter comes, and are less excited by going out and seeing people? Is it bad if for a period of the year our high powered jobs thrill us less, or if we become more contemplative and wonder at our place in the grand scheme of things? Rather than denying this process, might there not be some merit in embracing it, in seeing it as the natural cycle of seasonal changes, a kind of mental storing up of reserves in readiness for the following spring?

Writing in the 1960s, at a time when we all were being wowed with the promise of a leisure-filled future as technology removed the drudgery from our lives (of course we now all feel that we work longer than ever before), the philosopher Sebastian de Grazia commented (de Grazia, 1962):

Perhaps you can judge the inner health of a land by the capacity of its people to do nothing – to lie abed musing, to amble about aimlessly, to sit having coffee – because whoever can do nothing, letting his thoughts go where they may, must be all at a peace with himself.

This echoes the point of Thomas Szasz, one of the proponents of the anti-psychiatry school of thought, who would say to his students (Oliver, 2006):

'Has she got an illness called depression, or has she got a lot of problems and troubles which make her unhappy?' He turns and writes in large block letters: 'depression.' And underneath that: 'unhappy human being.' 'Tell me,' he says, facing the class, 'does the psychiatric term say more than the simple descriptive phrase? Does it do anything other than turn a "person" with problems into a "patient" with a sickness?'

Sufferers say that SAD is less severe than clinical depression, but it is a lot worse than just feeling a bit 'down', as perhaps the more extreme environmentalists would have it. SAD seems to hamper the quality of vocational, family and personal lives of those who suffer from it. Symptoms such as daytime fatigue and somnolence have significance in daily functioning, including occupational safety. Rosenthal himself suffered from SAD, and he wrote that in his first winter in New York he was reduced to just 'hanging in', as he put it, to 'try to keep everything afloat' (Rosenthal, 2006).

Websites devoted to SAD are littered with personal testimonies of the effects of the illness. The Catholic Church considers it serious enough to have designated Saint Pio as the patron saint of stress relief and what it calls the January blues. Rosenthal originally defined it as a syndrome in which depression developed during the autumn or winter and remitted in the spring or summer, but it is now classified as a form of recurrent depressive or bipolar disorder, characterised by episodes that vary in severity with a seasonal pattern (some people get a summer version). The *Diagnostic and Statistical Manual of Mental Disorders* (DSM) of the American Psychiatric Association, something of the bible of the profession, lists it as a subtype of affective disorder (mood disorder) with a seasonal pattern (Rodin & Martin, 1998). The DSM criteria include:

There must be a relationship between the onset of depression and a particular time of year.

Depressive symptoms must disappear at some other time of the year.

These onsets and remissions must have occurred at these times for at least the last 2 years.

Seasonal depressive episodes must outnumber non-seasonal depressive episodes over the person's lifetime.

These are still not completely clear-cut criteria, but that tends to be true in psychiatry, in which multiple conditions are not always clearly defined, including terms such as 'melancholia'. In practice, repeated season-linked changes in mood, energy, weight, duration of sleep, appetite and social activity are considered to make up winter SAD syndrome.

Nicholas Mrosovsky, the Canadian biologist and expert on adaptations of animals to seasonal change, subjected the early descriptions of SAD to rigorous analysis. He pointed out that alluring analogies to hibernation that gave the illness a biological rationale were misleading, because no mammal over about 5 kg in mass is a true hibernator. Larger mammals are able to lay down fat faster than they can metabolise it, whereas small species deplete their fat reserves in only a few days unless they truly hibernate.

He was also scathing about some of the early protocols that found people with alleged SAD by advertising for them. Mrosovsky wrote (Mrosovsky, 1988):

> Let me take a devil's advocate position to the problems of defining SAD by considering a spatial analogy. Suppose that the distribution of moles (skin blemishes) on the body is random or relatively so. We then define anyone with three or more moles in a row as having Mole Alignment Disorder (MAD). We then advertise for people who have this problem. After excluding some respondents (moles too small, not perfectly aligned) we have a sample. We can then study the physiological and psychological concomitants of MAD.

Powerful stuff, but despite this Mrosovsky concluded (Mrosovsky, 1988):

> One thing that emerges with absolute certainty from the study of seasonality in animals is the impressive capacity for change: from thin to fat, from dark coat to white, from sexually motivated to disinterested, from active to lethargic. The enormous and widespread capacity for seasonal change in mammalian physiology is unlikely either to be absent in our own species or to be unsusceptible to pathological manifestation.

In general, twenty-first-century medical science has abandoned many of the ancient notions of the importance of seasonal influence on health and disease. But it is not so easy to dismiss seasonal effects on humans, even though we are not certain that we are generally photoperiodic and many of us probably are not. Not only are depressive symptoms seasonally modulated, but even the response to placebo is. The ten-day response rate to placebo in double-blind controlled trials of various antidepressants carried out at the New York State Psychiatric Institute was analysed according to time of year. Threefold higher response rates occurred in summer than in winter (Wirz-Justice, 2005).

The incidence of SAD seems to increase with latitude, ranging from 1.4 per cent in Florida at 27° N, to 12.5 per cent at 40° N in New York (Rosen *et al.*, 1990), whereas, as expected, SAD was absent in the Philippines at a latitude of about 15° N. SAD seems to increase with latitude, but as with so many findings in biology, things are not that simple. Although winter SAD is presumed to be caused by lack of sunlight and so would be expected to show a higher incidence during the short winter days at the higher latitudes, a direct causal relationship has not yet been identified, nor is there a neat relationship between latitude and incidence. For example, there are low rates of SAD in Japan, with a reported prevalence of 0.89 per cent at 39.75° N and 0.48 per cent at 33.35° N.

Perhaps the most intriguing findings have been in Iceland, which lies on the edge of the Arctic Circle. Icelanders living at 64–67° N have an incidence of 3.6 per cent, in contrast with the 7.6 per cent among Americans living at much lower latitudes (Axelsson *et al.*, 2002). On the other hand, at

64° N in Siberia, the reported rate is 16.5 per cent (Axelsson *et al.*, 2002). It is of course relevant that although latitude correlates with the photoperiod, it does not equate with the weather in winter.

Canadians of wholly Icelandic descent living in Winnipeg, Manitoba, had a prevalence of SAD of 4.8 per cent, in contrast with 9.1 per cent among their neighbouring Canadians of non-Icelandic descent. It could be that the difference between Icelanders and other populations is down to genetic differences. Although the first Viking settlers inhabited Iceland only about 1,200 years ago, that is long enough for positive selection on genetic variation to work in humans. For example, about one-fifth of European women have an inversion of a piece of chromosome 17 that affects family size. Icelandic women possessing the inversion have 3.5 per cent more children than those without. Because they are contributing more children to succeeding generations, the variant is becoming more common (Stefansson *et al.*, 2005).

However, these inconsistencies in latitude and prevalence of SAD might just be because SAD is a more severe expression of the naturally occurring seasonal variation in mood experienced by a large percentage of the general population.

The best evidence that there is a recurrent, seasonally influenced, winter depressive condition that is at the very least a function of the interaction of light and internal circadian and/or circannual rhythms is the very high success rate of phototherapy. People feel lousy, receive light treatment and feel better. This happens too often to dismiss, although it says nothing about how light, or the lack of it, triggers SAD. But since Herb Kern was first treated, 85 per cent success rates have been claimed in patients reporting improvement in their wintertime condition after receiving phototherapy.

There is the long-standing issue of how the researchers can control for any placebo effects of light administration. Because there is a popular belief that SAD is due to the short winter days, SAD patients might consider their depression as being due to a lack of light. Light then serves as a conditioned stimulus, the response being elevation of mood. There are inherent difficulties in studying light therapies (Golden *et al.*, 2005):

> While it is relatively easy to create a placebo pill or capsule that is identical in appearance to an active medication formulation, it is more difficult to 'blind' a

subject when broad-spectrum intense white light is the active experimental intervention.

However, the basis for the claims for the efficacy of light therapy has recently been substantiated. In 2005 an expert group convened by the American Psychiatric Association reported that when they had analysed the data from all available randomised, controlled trials that met their *a priori* standards, there was (Golden *et al.*, 2005)

> a significant reduction in depression symptom severity following bright light therapy in seasonal affective disorder and in non-seasonal depression, as well as a significant effect with dawn simulation in seasonal affective disorder. In other words, when the 'noise' from unreliable studies is removed, the effects of light therapy are comparable to those found in many antidepressant pharmacotherapy trials.

It has taken some 25 years to reach this position, and the expert group has pointed out that the difficulties in getting the medical profession to accept phototherapy, despite what would seem to be its undoubted efficacy, may lie in the fact that light is not patentable and so its use as therapy will always be under-trialled and under-marketed relative to far more profitable pharmaceutical options. New drugs or other treatments cost a fortune to bring to market and achieve wide use among doctors. Without the funding and backing of pharmaceutical companies, it is hardly surprising that the expert group found that 'most of the published research reports on the effects of light therapy in mood disorders did not meet recognised criteria for rigorous clinical trial design.'

Despite this, light therapy has a real and positive beneficial effect in helping people suffering from affective disorders, particularly in those whose illness has a seasonal pattern. There is even a US patent for a mobile phone to be fitted with a number of light panels on the face of the phone which would deliver 'light therapy' to the user's face when the phone is in use. But this still does not help to explain the causes of the illness.

One of the first hypotheses about SAD was that the shorter winter photoperiod led to depressive symptoms. Therefore, exposure to bright light at the beginning and end of the winter day should simulate a summer photoperiod and restore summer behaviours. This was the rationale, as we have seen, for the first NIMH light therapy studies that used three hours of light exposure given at 6.00 a.m. to 9.00 a.m. and 4.00 p.m. to 7.00 p.m.. However, it was later found that photoperiod extension alone was not usually effective for SAD, and that single, intense daily pulses of light were as effective as the morning and evening pulses of photoperiod extension.

Nor was melatonin suppression alone enough to produce a therapeutic response. People with SAD whose melatonin levels were measured throughout the night in both winter and summer produced much higher levels in the winter than those who did not have SAD symptoms, but the summer levels were normal. However, when melatonin levels were suppressed, for instance with atenolol, a long-acting β-blocker used by millions of people with hypertension, there was no effect on SAD, so melatonin alone is unlikely to be the cause. However, there is some evidence that propranolol, a short-acting β-blocker, does have beneficial effects on SAD (Lam & Levitan, 2000). It may be that propranolol's shorter half-life minimises the amount of time for which melatonin secretion is inhibited, but the evidence is still ambiguous at best.

Another idea was that there is at the very least a genetic predisposition to SAD, as mentioned with regard to Icelanders. Genetic effects accounted for 29 per cent of the variance in seasonality (as assessed using a self-report questionnaire) in a study of 4,639 adult twin pairs in Australia (Madden *et al.*, 1996). Overall, genetic predisposition to seasonality was associated with so-called 'atypical' vegetative symptoms of depression, such as increased food intake, weight gain and increased sleep, compatible with treatment studies showing these symptoms to be the best predictors of a good response to light therapy.

Because several genes code for serotonin transport, and following on from the predisposition idea, an important research thread has been to investigate the main monoamine transmitters, such as serotonin, dopamine and noradrenaline, that have been implicated in mood disorders. Serotonin is of particular interest because lower-than-normal levels are correlated with clinical depression. Furthermore, the strong craving for sweets is the single physical symptom that seems to correlate best with SAD. If people with

SAD have difficulty in regulating serotonin levels during the winter, then their craving for carbohydrates is a way of compensating, because carbohydrates are believed to increase the level of the neurotransmitter. Fluoxetine (Prozac), the selective serotonin re-uptake inhibitor used widely as an antidepressant, is nearly as effective as phototherapy in treating the symptoms of SAD (Lam *et al.*, 2006). These are difficult studies in humans because certain neurotransmitters are more easily investigated than are others; for example, the risk of inducing psychosis or addiction greatly limits our ability to examine the dopamine system directly. Further, these neurotransmitters are functionally linked at many levels, making it difficult to sort out which neurotransmitter is implicated in what. So the causal mechanism is as yet unknown.

The association of SAD with dysfunction in the regulation of serotonin levels may explain why there is an overlap with other conditions, including generalised anxiety disorder, panic disorder, bulimia nervosa and chronic fatigue syndrome, that also have a seasonal component. SAD also may be associated with attention-deficit/hyperactivity disorder (ADHD). Both conditions have been described as 'disorders of central under-arousal coupled with a heightened sensitivity to stimuli from the physical environment' (Golden *et al.*, 2005), and both are more common in women.

Another theory puts alcohol in the frame. A seasonal pattern of alcohol use also may be associated with SAD. Some patients with alcoholism may be self-medicating an underlying depression with alcohol or manifesting a seasonal pattern to alcohol-induced depression. Such patterns seem to have a familial component and, like the link between ADHD and SAD, may be related to serotonergic functioning (Sher, 2004).

Although there may be several factors contributing to the development of SAD, the most likely cause results from a fundamental misalignment of the timing of the internal circadian rhythms that govern many of our daily activities. The idea is that in SAD, the timing between core rhythms such as cortisol levels and body temperature, as well as the timing of sleep with respect to the day–night cycle, are phase-delayed relative to the external clock or to other rhythms. This happens on a daily basis, as Wirz-Justice has described (Wirz-Justice, 2005):

The circadian component of mood follows the circadian rhythm of core body temperature rather closely. We wake up in not too good a mood, but this improves throughout the day to reach a maximum in the evening, and then mood declines during the night. The wake-dependent component reveals that we are quite cheery after a good night's sleep when sleep pressure is low, but that thereafter mood declines monotonically with time awake. If the temporal alignment between the sleep–wake cycle and the circadian pacemaker affects self-assessment of mood in healthy subjects, it might be expected that this is even more important for patients with depression. The phenomenon of diurnal mood variation as a characteristic of depressive state may indeed arise from phase relationships gone awry.

Any such misalignment brings with it the propensity for mood fluctuation, particularly in vulnerable individuals, and recent work suggests that it also happens with a seasonal pattern. The proposition is that later sunrises in the winter can delay a person's internal circadian rhythms so that they are internally desynchronised with their sleep–wake cycles and clock times. Providing this subtype with a corrective phase advance, such as early morning light, could act as an antidepressant for this group by realigning the endogenous core rhythms with the sleep–wake cycle. Similarly, the smaller subtype who are phase-advanced, perhaps cueing to the early winter dusk, would preferentially respond to a corrective phase delay from evening light.

Al Lewy and his colleagues at the University of Oregon tested this phase shift hypothesis in what may prove to be a seminal series of experiments. Their self-imposed task was to shift the circadian phase of a sample of SAD patients and measure this against an assessment of the depressive state of these individuals over a four-week period. The significance of what they achieved was reflected on by Lewy and the other authors: 'SAD may be the first psychiatric disorder in which statistically significant correlations are found between overall symptom severity and a physiological marker before, and in the course of, treatment in the same patients' (Lewy et al., 2006).

The devil in these experiments is in the detail, which is worth considering at some length, but in essence they had to (1) phase-shift the people they were examining, (2) measure the extent of the shift and (3) correlate this with an assessment of their depressive state.

Melatonin was the phase-shifting agent because it is thought to have effects on the phasing of the circadian system that are the opposite of light. Depending at what time it is administered, light can shift the circadian rhythm in either direction. Light in the first half of the night causes a delay in the activity rhythm during the following day, and conversely an advance if the light is administered in the second half of the night. Administering melatonin in the morning or evening to SAD patients to increase the duration of 'the biological night' would not be expected to be of any therapeutic benefit, unless it induced circadian phase shifts.

Melatonin had to be taken in the opposite half of the photoperiod to therapeutic bright light. Accordingly, most SAD patients (who are phase-delayed) should preferentially respond to afternoon or evening (p.m.) administration of melatonin (which would be expected to cause phase advances) compared with morning (a.m.) administration (which would be expected to cause phase delays). The experiment ran in the January/February period each year for four years; each year, one group of SAD patients took p.m. melatonin, another group took a.m. melatonin and a third group took a placebo (it was a different set of patients each year). All the patients kept sleep diaries, and in all years apart from the first they wore activity monitors on their wrists that recorded night-time activity and so measured sleep onset and duration.

The physiological marker that was measured was the dim-light melatonin onset (DLMO). This is the time when the evening rise in melatonin levels (sampled under conditions of dim light to avoid suppression of its production) continues above a certain threshold, operationally defined as 10 µg/ml in blood plasma. Comparing the time of DLMO with the sleep duration pattern gave a measure of the phase shift.

The results were clear. Most people with SAD are phase-delayed, and this circadian component of their condition can be shifted with beneficial therapeutic effect by appropriately timed administration of melatonin (Lewy et al., 2006).

SAD is a depressive condition that is seasonally triggered in people susceptible to clinical responses to a circadian phase shift. This depressive condition has a seasonal pattern that is a response to regular changes in the physical environment and not linked to seasonal psychosocial stressors such as holidays and ritual festivals. They are innate responses, not acquired (Harrison, 2004). The incidence of SAD is significantly higher in women than men, and it decreases with age. It is perhaps the most common mood disturbance in females of childbearing age, unremittingly experienced year after year during the six months between the autumnal and vernal equinoxes at temperate latitudes where there are marked seasonal changes in natural daylength. But people who get SAD can be treated with phototherapy, antidepressants and possibly melatonin, and also there have been claims for success with cognitive behaviour therapy.

Many of us get a little 'down' in winter, and in the past this may have been a useful protective measure as people slowed down, did less, used up less energy and so managed the harsh conditions of winter. Now we live in an artificially lit environment; some people are less sensitive to artificial light and need much more of it to synchronise their basic circadian rhythms effectively.

# 12

# THE SEASONALITY OF DYING

'We do not choose the day of our birth nor may we choose the day of our death, yet choice is the sovereign faculty of the mind.'

THORNTON WILDER (*THE EIGHTH DAY*, 1967)

Nearly 2,000 years ago the Roman philosopher Seneca wrote (Noyes, 1973):

Just as I shall select my ship when I am about to go on a voyage, or my house when I propose to take a residence, so I shall choose my death when I am about to depart from life.

He was wrong on that one. The emperor Nero accused his former teacher of plotting against him and ordered him to commit suicide. In what seems to have been one of the messiest demises in history, Seneca cut his wrists, took poison, jumped into a hot pool to increase the blood flow and finally died, according to Tacitus, from suffocation from the steam rising from the pool (Tacitus, 1971).

Although Seneca's thoughts on choosing the time of death were sheer hubris, the general idea that humans can 'give up' or 'let go' or 'hold on' as the end of life draws near is comforting to some or disconcerting to others according to taste. Notwithstanding a lack of evidence, there is a popular belief that people have some capability to choose the timing of their death and that there is a rise in death rates after a significant event such as a birthday or religious festival. American social scientists studying the 'anniversary reaction' analysed the deaths from natural causes around the birthday for 2,745,149 people (Phillips *et al.*, 1992). Women were found to be more likely to die in the week after their birthdays than in any other week of the year. In addition, the frequency of female deaths dipped below normal just before the birthday. The results did not seem to be due to seasonal fluctuations, misreporting on the death certificate, deferment of life-threatening surgery, or behavioural changes associated with the birthday. The explanation given for these findings was that females are able to prolong life briefly until they have reached a positive, symbolically meaningful occasion. Thus, the birthday seems to function as a 'lifeline' for some females. In contrast, male mortality peaks shortly *before* the birthday, suggesting that the birthday functions as a 'deadline' for males.

However, a careful analysis by Judith Skala and her colleagues at Washington University of this study and many others concluded (Skala & Freedland, 2004):

> The studies published to date have not convincingly established that death can be postponed through force of will or hastened by the loss of the desire to live. Modest effects have been found in some subgroups but not others, around some occasions but not others, and by some investigators but not others, and critics have raised methodological concerns about many of the studies. This research has also produced very little information about any specific biopsychosocial mechanisms that might account for temporal variations.

So another beautiful theory is slain by ugly facts. But although death may not be at a time of our choosing, doctors, psychiatrists and social scientists have been fascinated for many years by the role of the seasons in the timing of our deaths.

Many other behaviours such as indecent assaults, deliberate self harm and even suicides show a seasonal variation (Figure 12.1). Albert Leffingwell, an American doctor, published figures on suicides in European countries in his 1892 book entitled *Illegitimacy and the Influence of Seasons upon Conduct* (Leffingwell, 1892); it was five years later that Emile Durkheim's treatise on

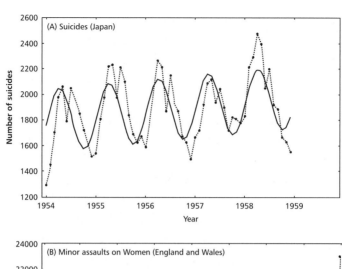

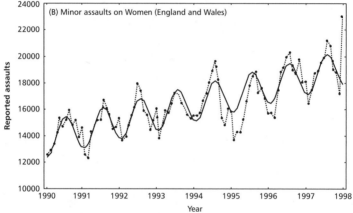

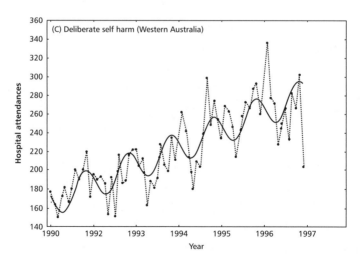

**Figure 12.1** Seasonal variation in various indicators of impulse control and inhibition. (A) Seasonal variations in the number of suicides (Japan). (B) Reported assaults on women (England and Wales). (C) Number of hospital attendances associated with deliberate self harm (Western Australia). In each graph the raw data (calendar-adjusted) are shown by the dotted line and the seasonal pattern by the solid line. Source: original data kindly supplied by Dr Daniel Rock, Centre for Clinical Research in Neuropsychiatry, The University of Western Australia.

suicide (published in 1897) provided the first serious efforts to combine sociological theory and empiricism to explain a social phenomenon. Durkheim collected the data on suicides in several European countries, and among his findings, which echoed Leffingwell's result, was a marked seasonal pattern. Suicide rates in the northern hemisphere peaked in May and June, or, as he put it, 'suicide reaches its maximum during the fine season, when nature is most smiling and the temperature mildest' (Durkheim, 1897). This seems odd, because intuitively most of us assume that suicide rates are higher in the dark and cold of winter rather than the sunny days of spring or early summer.

But his figures were correct then and they are still correct now. Although the Samaritans receive most calls around Christmas, historically the peak time for actual suicides, particularly violent suicide, is May and June in northern latitudes (Figure 12.1A) and November and December in the southern hemisphere. The amplitude between the peak and the trough of the suicide rate is about 20 per cent.

As befits the father of sociology, Durkheim's explanation for what may be thought of as the highly individualistic act of suicide was essentially social. He noted that suicide rates are connected to the length of day, but he believed that the suicide rate of a society was what he termed a social fact about that society.

He came to the view that the factors associated with higher numbers of suicides must relate to 'the time when social life is at its height'. The time of day, the day in the week, the season of the year, and so on, are not in themselves the reason for the changes in the number of suicides. Rather, the times when social life and interactions between people are greater are also those associated with increased suicide. And in the longer days there is more social interaction. In Durkheim's analysis, there is no causal relationship between suicide rate and amount of sunshine, but increasing daylength changes the social milieu, and it is that social milieu that had a causal relationship to suicide rate.

A biological explanation that was rejected by Durkheim was that it was the warmth of spring and summer that increased the suicide rate. The belief was that heat causes discomfort and irritation. But if heat were a factor, the suicide rate in hot countries should be higher than in colder ones, but the opposite tends to occur.

A third, and what can be termed psychological, reason relates to unmet expectations. We associate spring and, to a lesser extent, summer with new beginnings. People who are depressed may well make it through the bleak winter by looking forward to that new beginning. However, when they reach that expected point, namely spring, their life fails to change positively in the way they may have expected. The disparity of those expectations with reality may be the proverbial straw that breaks the camel's back.

While there may or may not be some truth in the above explanations, some researchers believe that there is sufficient evidence to suggest that a key biological factor in the seasonal variation in suicide is directly related to the variability in the hours of sunshine (Heerlein et al., 2006).

A research team based at Athens University analysed suicide statistics for the twenty OECD countries (eighteen of which are in the northern hemisphere). Their extensive studies strongly supported the early summer peak in suicide, and in their words 'the degree of seasonality, that is, the amplitude of the seasonal variation, can be largely explained in terms of monthly sunshine, both within and across countries' (Petridou et al., 2002).

If bright sunshine correlates with an increase in suicide, does this put a question mark against the bright light therapy for SAD? The short answer is no. Only 3 per cent of patients with SAD in an American study experienced a slight worsening of suicide scores after light therapy. Moreover, no patient attempted suicide or discontinued light therapy because of perceived elevated suicide risk (Lam et al., 2000). Depression is clearly not the same condition as SAD.

Sunshine generally seems to have a beneficial effect on mood, so it seems strange that bright sunlight in itself is correlated with increased suicide risk.

One hypothesis is that sunshine acts as a natural antidepressant but that first it improves motivation and then only later improves mood, thereby creating a potential short-term increased risk of suicide during the lag period. When the ten years' worth of Greek daily suicide and solar radiance data were carefully analysed, an increase in solar radiance during the day before the suicide event was significantly associated with an increased suicide risk. In fact, the average level of solar radiance during the four previous days was also significantly associated with an increased suicide risk. The increase in suicide risk in June compared with December, attributable to the daily sunshine effect, varies from 52 per cent to 88 per cent (Papadopoulos *et al.*, 2005).

Gavin Lambert of the Baker Heart Research Institute in Melbourne and his colleagues examined ten years of suicide statistics in Victoria, Australia (Lambert *et al.*, 2002). They found a marked seasonal variation that matched daylength. Suicides peaked in February and September and were at their lowest in June. He considers that this may point to the involvement of the neurotransmitter serotonin, because he has found a link between bright sunlight and brain levels of serotonin in healthy human subjects, and others have found serotonin disturbances in the brains and cerebrospinal fluids of suicide completers (Arehart-Treichel, 2003). It may be that the suicide has low levels of serotonin at the time of self harm. But how serotonin might trigger suicidal behaviour in certain people is far from clear.

Suicide in the terms in which we are speaking, which precludes suicide for political or religious reasons, is multifactorial and there is no single cause and effect. Further, suicide is an outcome of self-harming behaviour that results in death, and this relationship between self harm and suicide means that caution must be used in interpreting seasonal data because there may be a distinction between the timing of self harm and actual death.

However, the seasonality of suicide is a well-substantiated epidemiological characteristic, although there is some debate as to whether the variation is diminishing. Each year, some 800,000 to 1,000,000 people around the world deliberately kill themselves. As the World Health Organization has pointed out, that is about one person every 40 seconds. Suicide is the thirteenth leading cause of death worldwide and the third leading cause of death among those aged 15–44 years (Krug *et al.*, 2002). More people die from suicide than from road traffic accidents. For every suicide there are perhaps 20 attempts, many of which result in hospitalisation. The true numbers of suicide completers and attempted suicides are almost certainly substantially higher than recorded. It is a tricky area in which to get definitive data, and trend comparison is hard, but in the past half century and taking population increases into account, suicide rates may have increased by 60 per cent worldwide. Historically, suicide rates tended to rise with age, but in recent decades the rate among younger people has been rising and in many societies, contrary to the demographic changes, the absolute number of suicides in people under 45 years of age is higher than among those over 45 years old (CDC, 2005).

The seasonal variation in suicide may be declining. A group of Swiss researchers studied suicide rates going back 125 years. They found a smoothing-out of suicide seasonality in Switzerland that has been a continuous long-term process, which probably started by the end of the nineteenth century. Their conclusion is that seasonal effects in suicide will probably fade away in most regions of Switzerland in parallel with the disappearance of the traditional rural society. However, the changes were not homogeneous across all groups and the researchers themselves commented, 'the seasonal effects have been most impressive in drowning and hanging suicides, and in rural Catholic regions' (Ajdacic-Gross et al., 2005).

One explanation for the steep decline in seasonality of suicide in some countries is the successful prescribing of antidepressants. Hungarian researchers at the University of Budapest found a marked and significant seasonality in suicides, with a spring and summer peak between 1981 and 1989, when the prescription of antidepressants was relatively low and stable. However, this seasonality disappeared in the period between 1990 and 1996, when the prescription of antidepressants increased markedly, indicating that more and more depressed patients were treated pharmacologically (Rihmer et al., 1998). In contrast, the evidence from both the USA and Australia on the overall seasonality of suicides in those countries is as firm as it always has been.

The picture is confused because seasonal risk seems to vary separately with gender and also between violent and non-violent methods of suicide. In general terms, men choose more violent deaths than women. For instance, American men who commit suicide tend to shoot themselves, whereas women take pills. Psychopathology is a good indicator of potential suicide, and a Canadian study found that nearly all the schizophrenic males in the study who committed suicide did so in the autumn or winter. However, suicide peaks for males with different depressive disorders came at different times of the year, and in some disorders there was no seasonality (Kim *et al.*, 2004).

Suicide is a fascinating example of the complexity that arises in meshing large-scale epidemiological studies with small-scale genetics and molecular biology. And it keeps bringing up the earlier point made by John McGrath about gradients. If there is an hours-of-sunshine and suicide gradient, it provides a point of traction that may enable us to get at least one handle on this multifaceted, but vitally important, issue.

Apart from the seasonality of suicide, there is also a seasonal variance in violence in general. However, as with suicide, great care must be taken in sorting out all the confounding variables involved. Violence against other persons peaks in the mid or late summer months in England (Figure 12.1B) and a month or so later in the USA. Gunnar Morken and his colleagues found a late spring peak in Norway and a second peak in October and November (Morken & Linaker, 2000).

Daylength and sunlight may be key factors, but although there was no apparent temperature gradient with regard to the frequency of violence in the Norwegian studies, this is not of itself sufficient to dismiss the popular notion first enunciated some 170 years ago by the Belgian astronomer Adolphe de Quetelet that violent, inter-personal crime increased in the hotter summer months.

De Quetelet noted that crimes in various European countries varied by season and although violence increased in the summer, property crimes peaked in the colder months. He formulated these ideas into his thermic law of crime, which has been influential in criminological studies and the mainstay of the fiery Mediterranean temperament beloved in Grand Opera.

However, hours of daylight, or temperature, or the occurrence of a major sporting event are not the only factors in the incidence of violence. A rise in the real price of beer results in a fall in the rate of violence (Matthews *et al.*, 2006). This makes analysis of the seasonal factors very difficult because there are so many entangling variables.

Sociological explanations for the summer rise in violent crime are based on levels of social interaction. The more interaction there is, the explanation goes, the more violence occurs. Biological explanations tend to focus on seasonal variation in hormone and neurotransmitter levels. The answer is almost certainly a mixture of both.

Fortunately for the analyst, the same complexity and mix of factors are not quite true of the overall seasonality of death. In the UK, about 250 more people die each day during January than during August. There is a similar pattern in Canada and other countries in higher latitudes. Even in Australia's comparatively mild climate, there are more deaths in the winter months of June, July and August, and fewer deaths in the summer months of December, January and February. In 1999, each day in August averaged 400 deaths, whereas each day in February averaged 316 deaths (De Looper, 2002).

The seasonality of death is being influenced by our changing living standards. There has been an overall decline over the past couple of centuries in mortality from diseases that had previously been most severe in the summer months, such as dysentery, gastroenteritis, tuberculosis and other infectious and parasitic diseases. These declines are attributed to rising living standards, improved nutrition and food storage, cleaner water supplies and better sanitation (Taylor et al., 1998). Over the same period, however, there has been an increase in mortality from diseases associated with colder months, such as diseases of the circulatory system and respiratory diseases such as pneumonia and influenza. But even this seasonal distribution is altering as the excess of winter deaths is declining, not so much as a result of central heating, but rather because of socioeconomic progress resulting from increased prosperity and social welfare.

The average daily numbers of deaths due to diabetes mellitus, chronic liver disease and cirrhosis, and diseases of the urinary system, show statistically significant seasonality similar to that of cardiovascular diseases, with all of them peaking in January. These chronic diseases have multiple systemic effects that can compromise individual health and make them more vulnerable to pneumonia and influenza. However, there is no marked seasonality to cancer-related mortality, although there is a marked seasonal variance in the growth of tumours. Breast cancer, a disease that will be diagnosed in one of every nine living American women, is an example of the seasonal balance between a person and cancer. Several large studies, on each side of the equator, have demonstrated that the likelihood of the discovery of breast cancer is highest in spring, lowest in autumn and intermediate in winter and summer. These large studies were with culturally diverse populations living in very different climatic and geographic conditions. Because the average breast cancer takes many years to develop, these studies tell us that breast cancer size (which reflects both the growth of the tumour cells and the woman's defences against that growth) is growing more rapidly in springtime and less rapidly in other seasons. Other measures of breast cancer biology, including the average size of resected tumours, the pathologist's assessment of their microscopic aggressiveness, the number of lymph glands into which the cancer has spread, the concentration of hormone receptor molecules within the resected breast cancer cells and the overall survival of women with breast cancer, are each affected by season (Hrushesky, 1991).

It is highly speculative, but there could be a relationship between the seasonal changes in the mode of our immune response, from antibody-mediated immunity in the winter to cell-mediated immunity during the summer. The suggestion of a transition between them opens a 'window of vulnerability' to breast cancer in the spring.

Cold weather in itself, unless it is extreme, is not a direct killer. Relatively few people die from direct exposure to cold (hypothermia) during the winter months. Even in the UK, which has a poor record on excess winter deaths among older people, fewer than 1 per cent of the deaths are attributed to hypothermia itself (Collins, 1993). Paradoxically, it is a survival mechanism in people and other mammals that constricts blood vessels in cold weather, to conserve heat and maintain body temperature, that is a winter killer. With less room for blood to move in the constricted vessels, blood pressure rises – along with the risk of fatal heart attack and stroke, which peaks during winter.

Those countries with severe winters, such as Canada, Russia and Scandinavia, have a smaller percentage of excess deaths during winter than, for example, the UK and many Mediterranean countries. James Goodwin, who heads up Help the Aged's research activity in the UK and whose own academic research work was on the effect of cold on the cardiovascular system of older people, calls this 'the paradox of temperature. It is the total profile of cold exposure and associated behaviour that is the threat'

(Goodwin, 2007). People living in severe climatic zones protect themselves better.

This may be of particular relevance to the UK, which has especially high excess winter mortality among older people, particularly those over 75 years of age. The reasons are counter-intuitive and are another salutary reminder that seasonal effects in humans are confounded by social issues and can be very difficult to unravel.

There is little socioeconomic gradient with winter death in the UK. This is unexpected and is at odds with current notions of vulnerability from fuel poverty, but it seems to be a robust finding. Paul Wilkinson, who has a deserved reputation for his pioneering work on health inequalities, has pointed out (Wilkinson et al., 2004):

> Although lower socioeconomic groups have high mortality in absolute terms, it is not obvious that they should also have a high relative increase in deaths during winter months unless they are more exposed to the principal causes of it – specifically low ambient temperature. But people in lower socioeconomic groups do not on average have cooler homes than people in higher socioeconomic groups. This may reflect behavioural influences, but also the fact that housing association and local authority dwellings are often as well, or better, heated than owner occupied dwellings, reflecting the relatively recent construction of much social housing and efforts by local authorities to improve home energy efficiency.

Wilkinson goes on to quote other studies that placed more emphasis on personal behaviours and have argued that much excess winter mortality is related to exposure to cold from 'brief excursions outdoors rather than to low indoor temperatures' (Keatinge et al., 1989). Even five minutes' exposure to a temperature below 10°C can raise blood pressure.

The conclusion from this work is that because the risk of excess winter death is quite widely distributed in elderly people, it may be that initiatives that target only low-income households have limited effect. Goodwin points out (Goodwin, 2007) that whereas

> the poorest and most disadvantaged who cannot heat their homes in winter are greatly at risk in the winter, policies must incorporate the clear evidence that cold

weather has little respect for social class, income, or social status. It can kill regardless.

Attempts to counter excess winter deaths in older people in the UK through targeting 'fuel poverty' among low-income groups are well-intentioned, but they are not as effective as they could be. They need to be linked with attempts to modify what might be called high-risk behaviour, namely going outside with inadequate clothing and engaging in low-activity levels such as standing at a bus stop.

Although it is important to heat the home adequately, it is also very important that when older people go outdoors they dress in layers to conserve body heat, and it is important to wear a hat, scarf and gloves to minimise the amount of skin exposed, because increases in blood pressure do not require full-body exposure.

The potential 'failure' in public policy in the UK is worrying for an ageing society. Most winter 'excess' deaths are related to lowered immune function, resulting from the re-allocation of energy resources to defend body temperature, and hence a lower resistance to illnesses. During milder winters, resistance to illness is higher and the mortality rate is lower.

The oldest age groups are affected by extreme climatic conditions in general: cold weather in winter as well as hot spells during summer. This does not bode well as we move into a period of more extreme weather in previously temperate and densely populated parts of the world. On the other hand, warmer winters would help. In the UK, a 1°C increase in average daily temperature in January would result in about 30 fewer deaths each day and this would hold true for temperatures up to 3°C higher than the January norm. However, as the author of the study notes, 'this does not mean that we shall live longer in a warmer climate, rather that we may be more likely to end our lives in summer than is the case at present' (Subak, 1999).

It is not the average or normal weather situation that is the problem; we do poorly when certain weather thresholds are exceeded, such as in a heatwave. These can be short-term, but only several hours of unusually hot weather can lead to markedly increased death rates, upswings in hospital admissions and increases in the number of individuals suffering from mental stress and depression.

In many places, only the warmest 10–15 per cent of all days in summer have an impact on human mortality. Some of these increases in mortality are greater than 100 per cent and, for a large number of cities in the developing and developed world, heat is the greatest weather-related killer.

Although very high temperatures can kill older people in particular, it could be the variability in weather that is responsible for increases in heat-related illness and death rather than the intensity of the heat itself. American cities such as Chicago, St Louis and Philadelphia, despite the presence of air-conditioning in most homes and workplaces, have more heat-related deaths than many tropical cities in the developing world. People living in highly variable summer climates are ill adapted to extreme heat, mainly because it occurs irregularly. In contrast, people living in tropical regions cope with excessive heat through adaptations in lifestyle, physiological acclimation and the adoption of a particular mental approach. Cultural or social adjustments, including house design and general urban area structure, may also explain why heat-related mortality is often lower in hot climates with low variability (Kalkstein, 2000).

Age and gender have a key role in seasonal mortality. Among men, fluctuations in seasonal mortality increase with age, but in women susceptibility starts at later ages and is restricted to cold temperatures. Men show increases with age at younger ages, and older men become susceptible not only to cold but also to any unfavourable climatic conditions such as summer heat (Rau & Doblhammer, 2003).

This raises an ethical issue: How hard should we try to save the lives of elderly people endangered by climatic conditions? Demographers who have looked at this closely suggest two counterpoised opinions with the true effect lying somewhere in between (Rau & Doblhammer, 2003):

> either one assumes that the subject is in a frail condition and would have died anyway relatively soon or the other assumption would be that the individual does not differ in his/her robustness from the rest of the population. Saving a life in the latter case would be 'perfect repair' in terms of reliability engineering. If accidents and infectious diseases dominated the seasonal fluctuations, the overall effect could rather tend towards the 'perfect repair extreme'. Contrastingly, if chronic diseases mainly shaped the seasonal pattern, the effect on reducing

overall mortality would either be relatively small or it would require more efforts in medicine, public health and general living improvements to obtain the same effect as in the other case where relatively inexpensive interventions like vaccinations may result in remarkable improvements.

Or, to put it bluntly, if those who die as a result of swings in temperature were likely to have died soon anyway, how much scarce resource should be used to save them? This should keep the medical ethics committees busy for quite some time!

# 1 3

# WE ARE ALL PHENOLOGY
# FREAKS NOW

*Phenology* – from Greek 'phaino' (to show or appear) and 'logos' (to study). The scientific study of periodic biological phenomena, such as flowering and migration, in relation to climatic conditions.

The open-pit iron mine at Hamersley in Western Australia is so big it can be seen from space. The iron oxide strata were laid down between 2.5 and 1.9 billion years ago. Although iron had already been spewing forth from the earth's molten interior in underwater volcanic eruptions for some millions of years beforehand, in the reducing environment of the time the iron remained in the water as the soluble ferrous form. But by 2.5 billion years ago, tiny photosynthesising cyanobacteria in shallow waters were producing enough oxygen to oxidise (rust) the iron dissolved in the seas, which then sank to the ocean floor as the insoluble ferric form. The length of daylight and the water temperature changed with the seasons and the photosynthesising cyanobacteria produced more dissolved oxygen in the summer and so more iron oxide was deposited. Some of this wound up at what is now Hamersley, where these seasonal periodicities are revealed by close analysis.

The evidence of the annual seasonal changes in climate is written in the rocks, trees, ice and coral that surround us. The Hamersley rocks record millions of years. Trees count in the hundreds and thousands. A section of a giant sequoia on display in London's Natural History museum comes from a tree that was well over 1,300 years old when it was felled in 1892. It had lived through most of our modern history from the time of Charlemagne to Gladstone.

In each of those 1,300 years, as summer drew to a close, the photoperiodic mechanism in the sequoia decoded the seasonal information of the shortening days that signalled that it was time to begin preparing for winter. Like many trees in temperate zones, the sequoia made one growth ring each year, with the newest adjacent to the bark. For the entire period of the tree's life, a year-by-year record was formed that reflected the climatic conditions in which the tree grew. A wet summer and a long growing season resulted in a wide ring; a drought year resulted in a very narrow one. By matching tree rings of known years to rainfall records, we can deduce how much rain each size ring represents. Depending on which part of the world the tree grew, researchers can even 'see' El Niño effects in trees that show signs of much rainier or much drier seasons than normal.

By studying rocks, trees, glacial strata known as varves, ocean sediments, ice layers and coral formations and allying them with radiometric techniques, we have developed a picture of the variation in seasonal climate in the past resulting from overall changes in the climate. While there are still some naysayers, the overwhelming consensus in the scientific community is that the planet is rapidly heating up again and the climate is changing, but this time it is probably because of human activity. As Lesley Hughes of Macquarie University has pointed out, 'advances in phenology in insects, birds, amphibians and plants provide the most compelling evidence yet for the impact of human-induced climate change, simply because alternative explanations are generally less plausible' (Hughes, 2000).

A possible rise in average global temperature of perhaps 3°C over this century will mean dramatic changes in local climate and weather. Rises in sea level will flood coastal areas, some deserts may bloom again and rich farmlands may become arid wastes. Harvests of maize, rice, soybean and wheat will diminish as increasing temperatures, drought and ground-level ozone concentrations will result in substantial decreases in crop yields, outweighing the beneficial fertilisation effects currently predicted from rising levels of atmospheric carbon dioxide (Slingo *et al.*, 2005). But no matter how hot it gets, the earth will still rotate every 24 hours and it will still orbit the sun once a year. The photoperiod will be the same as it has been.

What will change will be both the nature and the duration of the seasons, and it is this that will have an impact on plants and animals. Camille Parmesan, a biologist at the University of Texas, and Gary Yohe, an economics professor at Wesleyan University, looked at the reported impact of climate change on more than 1,700 species over a 20-year period. They concluded that, however you cut the data, it is likely that climate change will affect maybe 50 per cent or more of all plant and animal species in the wild over the next century or so (Parmesan & Yohe, 2003).

Although the tree rings and the rock strata tell us much about seasonal change in the past, they do not record the fine detail such as how the specific timing of the seasons has changed. There is no way of knowing from them whether spring was early or late in a given year and by how much. For that, we have to rely on more recent data.

Japanese priests and scholars have been recording the flowering of the cherry blossom for about 1,200 years, but the fathers of phenology proper were two eighteenth-century figures, the great Swedish botanist Carolus Linnaeus and Robert Marsham, an English landowner. Linnaeus, who is best known for inventing the classification system of species used universally by biologists, systematically recorded flowering times for 18 locations in Sweden over many years. His meticulous notes also recorded the exact climatic conditions when flowering occurred.

Marsham's interest stemmed from a desire to improve the timber production on his estate. He corresponded with Gilbert White, the author of the classic *Natural History of Selborne*, and although the two never met, they encouraged each other's interests. White was particularly interested in birds, recording the arrival of summer migrants and spring and autumn passages. But it was Marsham who began their systematic study. He began recording his *Indications of Spring* in 1736 on his family estate near Norwich in Norfolk and continued to note down significant dates for the next 62 years, recording some 27 natural events for more than 20 animals and plants. These included tree leafing times and the arrival of migrant birds. After his death in 1798, successive generations kept up this information gathering for 160 years until the death of Mary Marsham in 1958.

Over the past 300 or so years, the pioneering efforts of Marsham, White, William Markwick and others have been followed by a host of dedicated and even obsessive observers, many of whom were amateurs in the best traditions of natural historians. They have been logging all the firsts: the first appearance of frogspawn, the first flowering of daffodils, the first call of the cuckoo, the first sighting of swallows, bud burst in oaks, egg-laying in birds, and so on. And they have done the same, if not quite in reverse, at the other end of the year as they recorded the leaving of the birds in the autumn migration and all the lasts such as the final flowering of plants, and they have detailed the timing of the natural world at various points in between.

Many countries now have phenology networks to encourage the collection and collation of the information faithfully recorded by the members. Richard Fitter kept a record of the flowering times of hundreds of plant species around the Oxford area over nearly 50 years. Analysing the data, his son Alistair compared the flowering dates for the entire decade of the 1990s with those for the previous four decades. He found that 385 plants were flowering an average of 4.5 days earlier. Among the most dramatic changes was in the white dead-nettle (*Lamium album*), which now flowers 55 days earlier than it did: typically on 23 January in the 1990s compared with 18 March in the 1950s (Fitter & Fitter, 2002). The advanced flowering of this and nine other species is shown in Table 13.1.

The story is similar all over the northern hemisphere. Jerram Brown, now a professor emeritus at the State University of New York, Albany carefully logged the egg-laying behaviour of Mexican jays (*Aphelocoma ultramarina*) in the Chiricahua Mountains of southern Arizona for more than 30 years. In a now famous study, he showed that the jays are laying their eggs earlier and earlier each season. By 1998, the first eggs of the season arrived ten days earlier than in 1971 (Wuethrich, 2000). In the southwest Yukon, a University of Alberta biologist, Stan Boutin, has been studying the DNA and mating habits of female red squirrels for more than ten years. His team found that as spring came earlier, so did the squirrels' litters. They now give birth three weeks earlier than in previous records (Berteaux *et al.*, 2004). As we saw earlier, Marcel Visser and his colleagues have shown the mismatch between the timing of oak tree bud burst, winter moth caterpillar biomass and great tit egg-laying and hatching in the forests around Arnhem in

| FLOWER SPECIES | ADVANCE IN FLOWERING (DAYS) |
|---|---|
| White dead-nettle (*Lamium album*) | 55 |
| Ivy-leaved toadflax (*Cymbalaria muralis*) | 35 |
| Hornbeam (*Carpinus betulus*) | 27 |
| Spurge laurel (*Daphne laureola*) | 26 |
| Hairy bittercress (*Cardamine hirsuta* L.) | 25 |
| Lesser periwinkle (*Vinca minor*) | 25 |
| Greater stitchwort (*Stellaria holostea*) | 25 |
| Herb bennet (*Geum urbanum*) | 22 |
| Lesser celandine (*Ranunculus ficaria* L.) | 21 |
| Opium poppy (*Papaver somniferum*) | 20 |

**Table 13.1** The first flowering dates of ten plants from a total of 385 wildflower species studied by Richard and Alistair Fitter, who found that, on average, they bloomed four and a half days earlier during the 1990s than they did in the 1950s. These 10 species showed the largest advance in flowering (Fitter & Fitter, 2002).

Holland. Blue tits (*Parus caeruleus*) breeding in Corsica and southern France have had to cope with an advancing spring leaf flush and ensuing food availability that results in a mismatch between the timing of peak food supply and nestling demand, so shifting the optimal time for reproduction in the birds (Thomas *et al.*, 2001).

Land birds in the southern hemisphere are behaving similarly to their counterparts in the northern hemisphere, with advances in the timing of breeding and migratory arrival. However, there are few similar long-term biological time series to study. French biologists Christophe Barbraud and Henri Weimerskirch found a contradictory trend towards later spring arrival and laying when they examined a 55-year record of arrival and time of egg-laying of nine seabird species kept by generations of scientists at a research station on Adélie Land in eastern Antarctica, at a latitude of about 67° S (Barbraud & Weimerskirch, 2006). Species now arrive at Adélie Land 9.1 days later, on average, and lay eggs an average of 2.1 days later than in the early 1950s. However, Barbraud and Weimerskirch were studying records of seabirds that are the top predators in a marine food chain. In eastern

Antarctica it seems that the krill population, which is the basis of the marine food chain, is in decline. Seabirds suffer disproportionately from a reduction in food abundance at the bottom of the chain. Consequently, there is micro-evolutionary selection pressure that is favouring the birds arriving a few days later and compressing the pre-egg-laying process of territory selection and courtship by seven days. This shift may enable some of the bird species to survive in a declining food environment.

The complex changes occurring in the world's climate system are having different effects in different places. Because there is far more water than land in the southern hemisphere than in the northern, the effects of climate change will be different because the heat capacity of water has a considerable moderating effect on the weather. There will be different geographic responses to climate change, but this will not negate the general effect.

William Bradshaw and Christina Holzapfel have been working with *Wyeomyia smithii*, a North American mosquito that lives in a range that extends from Florida to the Canadian border. The mosquito lives within the water-filled leaves of the purple pitcher plant. In its specialised, stable developmental micro-habitat, the species relies exclusively on photoperiod for its seasonal cues, unlike other insects which may use a variety of cues including temperature and food availability, sometimes in combination with a circannual rhythm (Bradshaw & Holzapfel, 2001). During the late summer, individuals sense the shortening days and switch from active development and reproduction to diapause at a critical photoperiod. Bradshaw and Holzapfel reasoned that as a consequence of climate change, populations of mosquitoes in the northern states should show a more southern phenotype now than they did a few decades ago if they have been adapting to longer growing seasons and later onsets of winter. There should be progressively shorter critical photoperiods now than in the recent past.

Because Bradshaw and Holzapfel have been studying these mosquitoes for most of their academic careers, they were able to compare studies in 1972 with those in 1988 and 1996. They found a significant, progressive shift at increasingly northern latitudes towards shorter critical photoperiods at the later date. At 50° N, the critical photoperiod declined from 15.79 to 15.19 hours from 1972 to 1996, corresponding to nine days later in the autumn of 1996 than in autumn 1972.

Because the phenological timing of the developmental course of *Wyeomyia smithii* is wholly dependent on photoperiod, Bradshaw and Holzapfel could answer a key question with regard to the species: Do altered seasonal interactions result entirely from temperature-sensitive responses to the environment by individuals as expressions of plastic phenotypes, rather than as actual genetic changes in populations? Because there were no cues other than photoperiod, it was clear in *Wyeomyia smithii* that, with longer growing seasons, the mosquitoes have shown a genetic shift towards the use of shorter, effectively more southern daylengths, to cue the initiation of diapause.

Further, when Bradshaw and Holzapfel experimentally transplanted northern mosquitoes to a simulated southern climate, there was a huge loss of reproductive fitness, nearly all of which was due to experiencing the incorrect seasonal cues (photoperiod), whereas the warmer summer temperature of the more southern locality was not a factor.

The implication in this case is that as a result of climate change it is not that the summers are getting particularly hotter that is resulting in changes, but that the winters are not as cold and spring comes earlier (Bradshaw & Holzapfel, 2006):

> Northern populations experience shorter growing seasons than southern populations. For example, mean daily temperatures are above 10 ° C all year at 30 ° N but are above 10 ° C for only 2.5 months at 50 ° N. Climate change is proceeding fastest at the most northern latitudes, where the gradient in winter cold is steepest, thereby expanding the growing season while alleviating winter cold stress without imposing summer heat stress. Hence, northern climates are becoming more like those in the south. At least within insect species, northern populations use longer daylengths to cue the initiation of dormancy earlier in the autumn than do southern populations.

Those mosquitoes that responded to a shortened critical photoperiod and remained active until later in the autumn were able to gain a selective advantage, possibly because they could store a few more days' resources for the coming winter.

The survival value of a given critical daylength as the trigger for life course events is because photoperiod is the most reliable predictor of seasonal change (as we have been at pains to describe), so that many insects living in seasonal environments have evolved a means of sensing this cue. This is especially true for terrestrial habitats in which brief periods of unseasonable warming or cooling could send a false signal about time in the season. By contrast, animals living in large lakes and the oceans are buffered from short-term temperature fluctuations by the thermal inertia of water. The temperature of the medium in which they live is a more reliable indicator of the passing of the seasons than it is on land. Consequently, marine invertebrates such as copepods (small crustaceans), which in some respects can be regarded as marine insect equivalents, have a significant temperature modification of their response to photoperiod with regard to diapause.

The selection pressure on *Wyeomyia smithii* has resulted in a shift in the mean value of a fixed character, in this case critical photoperiod, in each individual in the population. The same genetic rationale for adapting to changes in seasonal timing may or may not be the case in other species.

Although many of the aspects of phenological behaviour have a heritable component, it would be wrong to underestimate the flexibility that exists within species, and indeed among different individuals within a species. Ian Newton has pointed out (Newton, 2008):

> Many of the changes observed in bird migratory behaviour need entail no genetic change, for in every aspect of migration there is scope for individual flexibility through which individuals can adjust their migratory behaviour to some extent according to prevailing conditions. For example, the same birds may arrive on their breeding areas earlier in warm springs than in cold ones, or they might migrate farther in cold winters than in mild ones, in response to differing food supplies. Hence as climate changes from year to year, or over longer periods, birds have considerable scope for adjusting their behaviour to match these changes, without the need for any modification in the genetic control mechanisms.

Many animals and plants live in environments in which the predictability and the amplitude of the environmental fluctuations vary. For instance, birds that live in habitats in which environmental cues such as photoperiod are poor predictors of seasons (such as equatorial residents, or migrants to equatorial or tropical latitudes) rely more on their endogenous clocks than birds living in environments that show a tight correlation between photoperiod and seasonal events. Many animals also occupy differing geographical ranges over the course of their lives. Accordingly, free-living vertebrates, and birds in particular, differ greatly in the timing and sequence of their life-cycle stages (Wikelski *et al.*, 2008).

Obligate bird migrants are overwhelmingly long-distance travellers whose migrations tend to be more or less at the same time each year and to the same place. Their timing mechanism is under genetic control. They leave their northern breeding grounds often well before food reserves are depleted. In contrast, facultative migrants may migrate in some years but not in others. Whether or not they travel is in response to prevailing local conditions, in particular food availability.

Birds do not fit into a polarised obligate/facultative migration dichotomy. Rather, there is a spectrum of behaviour from inflexible to flexible and there is some dynamism within this spectrum, with some species moving from one type of behaviour even within a single migration. A bird may start out in obligate mode but then switch into facultative mode with time and distance and stop when local conditions are favourable.

It is more than 70 years since Erwin Bünning put forward his hypothesis suggesting that circadian timing could interact with photoperiod to provide the means for plants and animals to adapt their behaviours, physiology and even anatomy to predictable seasonal changes. Since then, evidence has been found for endogenous circannual clocks in many animal species that essentially set an annual timing strategy that may be tuned by tactical proximal factors such as photoperiod, temperature, resource availability and rainfall. There is also evidence, as we explained before, that insects may use a different timing process to determine the timing of life events.

Bünning's hypothesis is still the overarching paradigm, but in the past 70 years we have learned how diverse and flexible plants and animals in particular are with regard to coping with seasonal change and how a suite of complementing mechanisms may be used. This flexibility is part of a general appreciation that nature is both more subtle and sophisticated than envisaged. At a molecular level the emerging picture is that the old mechanistic model of 'one gene – one protein' is old hat. There are some 30,000 genes in the human genome, but perhaps six or even more times as many proteins in the proteome as a result of the many ways in which the coding sections of DNA can be spliced together.

The genes themselves are turned on or off at different times in different cells. The timing of the switching of genes on and off largely determines development and hence the ultimate fitness of the adult. The control elements are products of gene expression; they can be complex, with feedback loops inhibiting the activation of a gene that might otherwise be undergoing activation. It is not so much that each gene determines a factor in the phenotype, although there are many such structural genes, but also that many genes additionally make a regulatory contribution through their proteins that is now recognised as being of fundamental biological importance because it means that there is potentially immense diversity and flexibility in the system.

The result is that organisms may have far more plasticity than was thought. Botanists have known for years that a cutting grown in moist conditions can look very different from a genetically identical cutting grown in dry soil. The simplifying assumption of evolutionary theory may be that genetic differences determine the relative success or failure of organisms in a given environment. But in practice, as was explained in the journal *Nature* (Dusheck, 2002), many organisms can respond with considerable flexibility to a changing environment. The idea of phenotypic plasticity – the ability of a single set of genes to generate a range of characteristics, or phenotypes, depending on the environment in which the developing organism finds itself – has spawned new attempts to study how organisms evolve in the context of their physical environments and of the other living things with which they interact. It adds ecology to evolutionary biology, developmental biology and genetics.

Those organisms that succeed will be those that have the flexibility to change and adapt to the new temporal regime as fast as the speed of the changes in seasonal climate timing. The pitcher-plant mosquitoes showed a marked change within five years, it took less than seven years for the necessary genetic modification that enabled field mustard plants in California to become better able to resist drought, and moderate changes were detectable over ten years in red squirrels. But in great tits, even after 30 years, only the portion of the population that is most able to modify the timing of egg-laying in response to earlier springs seems to have changed genetically.

Some species may override their photoperiodic cues and migrate pole-ward to remain in climatic regimes to which they are phenologically well adapted. Others may move to higher altitudes. But complete communities will not simply move north or south in synchrony with changing tempera-tures. Instead the blend of plants and animals will change. Climate change could create ecosystems that are unknown today.

Species that stay in place may survive by adjusting to changing condi-tions. Some, such as Bradshaw and Holzapfel's mosquitoes, will survive through genetic modification of the critical photoperiod. Other species may depend on their phenotypic plasticity, which enables individuals to cope with wider environmental stressors including seasonal climate timing. The great tits in Arnhem are failing to remain in synchrony with winter moth caterpillars, but birds of the same species at a research site near Oxford, England, some 400 km away, have altered life cycles in step with the same insects.

It is impossible to overstate how important the changes in seasonal cli-mate timing will be. There are feedbacks between phenology and factors affecting the global carbon balance that may act to accelerate or decelerate rates of climate change. Modest changes in the length of the season when plants are actively photosynthesising can significantly alter the storage of carbon. A forest at high latitudes can be a net sink or a net source of atmos-pheric carbon dioxide by virtue of a week more or less in the length of the growing season (Lechowicz, 2001).

Clocks and rhythms were a somewhat arcane subject in biology until the 1980s. In a recent flurry of activity, scientists around the world discov-ered the ubiquity and centrality of timing processes, and the synchronisa-tion by circadian and circannual rhythms with the external environment was recognised to be of profound importance. These anticipatory mecha-nisms have helped to determine the chronology of the life cycle in nearly all organisms and enabled them to mesh with the daily and seasonal changes in the environment.

Understanding the ways in which organisms have adapted and will adapt to seasonal climatic change will help us in the future in both mitigating and managing some of the effects of global climate change. On the basis of their studies, Bradshaw and Holzapfel predict that small animals with short life cycles and large population sizes will probably adapt to longer growing seasons and be able to persist; however, populations of many large animals with longer life cycles and smaller population sizes will experience a decline in population size or be replaced by more southern species (Bradshaw & Holzapfel, 2006).

A better understanding of the biological processes will help us to develop new agricultural and horticultural practices and to devise means to protect human health from attacks by both old and resurgent and new and insurgent pathogens, and will enable us to try to preserve and conserve other species. We will not save them all, and it is probably already too late to save many. Much of the awe and wonder that is the diversity of life on earth that we know at present will be lost.

As we have tried to show in this book, the exquisite temporal sensitivities of living organisms, established over countless generations, that enable them to predict the regular rhythms in the environment and so synchronise their life histories to maximise reproduction, are being split asunder. We are destroying not only the spatial markers of our world but also the temporal ones. And once gone, they, like the migrating birds, may not return at the same time next year.

# GLOSSARY OF COMMON TERMS

**Aestivation.** Adaptations, usually including a reduced metabolic rate, to cope with hot weather conditions.

**Amplitude.** (1) Difference between maximum (or minimum) and mean value in a sinusoidal oscillation. (2) Difference between maximum and minimum value of a biological oscillation.

**Autumn equinox.** The time in autumn when day and night are each 12 hours long and the sun is at the midpoint of the sky.

**Biological clocks.** Self-sustained oscillators that generate biological rhythms in the absence of external periodic input (for example, at the gene level in individual cells).

**Chromophore.** The light-absorbing molecule in a photopigment complex. The chromophore of animal photopigments is a specific form of vitamin A (11-*cis* retinaldehyde) that is bound to a protein called an opsin.

**Chronobiology.** Derived from the Greek (*chronos* for 'time', *bios* for 'life', and *logos* for 'study'), the word is used to denote the study of biological rhythms.

**Chronopharmacology.** The practice of administering medicinal drugs in time schedules that optimise the therapeutic action of the drugs (in consonance with the patient's circadian rhythms).

**Chronotherapy.** Use of treatment timed according to the stages in the sensitivity–resistance cycles of target (or non-target) tissues and organs (or of the organism as a whole) to enhance the desired pharmacological effect and/or to reduce undesirable side-effects of drugs or other therapeutic agents.

**Circadian rhythm.** A biological rhythm that persists under constant conditions with a period length of around 24 hours. From the Latin *circa* and *diem*, 'about a day'.

**Circadian time (CT).** The subjective internal time of an organism under constant conditions. By convention, CT 12 corresponds to activity onset for a nocturnal species, whereas CT 0 designates activity onset for a diurnal species.

**Circannual.** A rhythm with a period of about 1 year (±2 months), synchronised with or desynchronised from the calendar year.

**Clock gene.** Gene involved as a component of the molecular mechanism that produces a circadian oscillation.

**Clock-controlled gene (CCG).** A gene whose expression is regulated directly by the core oscillator mechanism.

**Cyanobacteria.** Photosynthetic bacteria, sometimes called 'blue-green algae'.

**Dampened oscillation.** Oscillation decreasing (dampened) in amplitude as a result of inevitable loss of energy.

**D:D.** Abbreviation for a lighting regime of constant darkness.

**Desynchronisation.** (1) External: loss of synchronisation between rhythm and zeitgeber. (2) Internal: loss of synchronisation between two or more rhythms within an organism.

**Diapause.** A term describing a state of low metabolic activity, usually in insects, associated with an increased resistance to environmental extremes and altered or reduced behavioural activity. Diapause occurs during one or more genetically determined stages of metamorphosis in a species-specific manner.

**Diurnal.** An activity or process that occurs during the daytime (light).

**E-box.** A nucleotide motif, CACGTG, involved in enhancing the expression of many clock genes.

**Eclosion.** Hatching of an adult insect from its pupal case.

**Entrainment.** The process by which a biological oscillator is synchronised to an environmental rhythm such as the light/dark cycle.

**Equinox.** On a day that has an equinox, the centre of the sun spends a nearly equal amount of time above and below the horizon at every location on earth, and night and day will be of nearly the same length. The word *equinox* derives from the Latin words *aequus* (equal) and *nox* (night).

**Enzootic.** A disease that is constantly present in an animal population but usually affecting only a small number of animals at any one time.

**Free-running.** The endogenous rhythm exhibited by a circadian system under constant conditions.

**Gene.** The definition of a gene is a subject of considerable debate among molecular biologists, partly because of the complexities of regulation and transcription. One definition that avoids this is that a gene is a union of genomic sequences encoding a coherent set of potentially overlapping functional products.

**Gene expression.** The process of making the product of a gene via transcription.

**Genome.** The genome of an organism contains all of the biological information needed to build and maintain a living example of that organism encoded in the DNA (or, for some viruses, RNA). The genome is a sequence of nucleotides.

**Genotype.** The gene make-up of an organism; the particular set of genes it possesses, even if not expressed.

**Hibernation.** Describes a state of inactivity and metabolic depression characterised by lower body temperature, slower breathing and lower metabolic rate. Hibernation may last several days or months depending on species, ambient temperature, and time of year. The typical winter season for a hibernator is characterised by periods of hibernation interrupted by sporadic euthermic arousals, in which body temperature is restored to near normal values.

**Insolation.** A measure of solar radiation energy received on a given surface area in a given time. It is commonly expressed as average irradiance in watts per square metre ($W/m^2$).

**Jet-lag.** A malaise resulting from a sudden move to a different time zone (often by a trans-meridian flight).

**L:D.** Abbreviation for a lighting regime consisting of alternating periods of light and dark.

**L:L.** Abbreviation for a lighting regime consisting of constant light.

**Masking.** The phenomenon whereby an external factor directly affects the expression of an overt rhythm.

**Migration.** A term used to describe a broad range of movements ranging from the small-scale passage of micro-organisms through the soil, the large daily vertical movements of plankton in the ocean, and long-distance foraging trips. In this book we use the term to describe behaviours that show a seasonal movement of populations between regions where conditions are alternately favourable or unfavourable, including one region in which breeding occurs. Animals migrate seasonally when they benefit in terms of survival and reproduction.

**Neuropeptide.** A protein hormone or messenger that is released from a nerve cell (neurosecretory cell) into the blood or intercellular spaces and changes the activity of a target cell. The surface of the target cell has specific receptors for a particular neuropeptide.

*Neurospora.* *Neurospora crassa* (bread mould). A filamentous fungus, used as a genetic model.

**Neurotransmitter.** A chemical messenger that is released from a nerve cell at its synaptic terminal and stimulates or inhibits a postsynaptic neurone by altering its electrical potential.

**Opsin.** The protein component of animal photopigments. Opsins use a vitamin A chromophore and possess a characteristic 'bell-shaped' absorbance spectrum.

**Oscillator.** A system capable of producing a regular fluctuation of an output around a mean. In chronobiology, an oscillator refers to the molecular mechanism within a cell capable of generating self-sustained rhythms.

**Overt rhythm.** An observable rhythm that is directly or indirectly regulated by the circadian clock.

**Pacemaker.** Structure capable of sustaining its own oscillations and of regulating other oscillators.

**PAS domain.** Protein sequence motif found in many clock proteins, involved in signalling pathways that transmit environmental information such as oxygen, redox state and light. Often associated with protein–protein interactions.

**Period.** The time after which a defined phase of an oscillation recurs.

**Peripheral oscillator.** An oscillator found in a tissue that is capable of regulating local physiology but is dependent on a pacemaker for entrainment. Also called a 'slave' oscillator.

**Phase.** A particular reference or reference point within the cycle of a rhythm (for example, onset of activity).

**Phase shift.** A single, persistent change in phase brought about by the action of a zeitgeber.

**Phenotype.** The observable characteristics of an organism; actual observed properties, such as morphology, development or behaviour.

**Phenotypic plasticity.** The ability of an organism with a given genotype to change its phenotype in response to changes in the environment.

**Photoentrainment.** The entrainment of an oscillator by the light/dark cycle.

**Photoperiod.** The duration of light and dark in a 24-hour, or near-24-hour, cycle.

**Photopigment.** A molecule that is capable of transducing the absorbance of a photon into an intracellular response.

**Pineal gland.** Neuroendocrine gland found in all vertebrates that synthesises melatonin. It is directly photosensitive in all non-mammalian vertebrates.

**Proteome.** The entire complement of proteins expressed in an organism.

**SCN.** Suprachiasmatic nuclei. Paired nuclei within the ventral hypothalamus that function as the circadian pacemaker in mammals.

**Solstice.** Technically, when the tilt of the earth's axis is oriented directly towards or away from the sun. The summer solstice is the longest day of the year and the winter solstice is the shortest.

**Spring equinox.** The time in spring when day and night are each 12 hours long and the sun is at the midpoint of the sky – 21 March.

**State variable.** A term used in mathematical modelling of oscillators, referring to a quantity that changes with time. More generally, a term used to denote a variable that is essential for defining the state of a system. Sometimes applied to the essential components (genes, proteins) required to generate a circadian oscillator.

**Summer solstice.** The longest day of the year in the northern hemisphere, when the sun is at its most northerly point in the sky – 21 June.

**Tau ($\tau$).** The natural period of a free-running biological rhythm.

**Temperature compensation.** A characteristic of circadian rhythms, whereby changes in temperature do not significantly alter the period.

**Torpor.** Usually a short-term state of decreased physiological activity in an animal, characterised by a reduced body temperature and rate of metabolism. Animals such as hummingbirds and many bats show a daily torpor with reduced temperature and metabolic rate during their rest phase.

**Transcription.** The process by which the information coded in a DNA sequence is transcribed, or copied, from DNA to RNA.

**Transduction.** A process that converts one form of signal to another, such as an external light signal into a physiological stimulus.

**Translation.** The process in which the genetic information encoded as a ribonucleotide sequence in mRNA is carried from the chromosomes to the ribosomes and is 'read' by translational machinery in a sequence of nucleotide triplets called codons. Each of those triplets codes for a specific amino acid.

**Transpiration.** The evaporation of water from the aerial parts of plants, especially the leaves but also the stem, flowers and roots. Leaf transpiration occurs through stomata and can be thought of as a necessary cost associated with the opening of stomata to allow the diffusion of carbon dioxide from the air for photosynthesis.

**Ultradian rhythm.** A biological rhythm with a period much shorter (that is, a frequency much higher) than that of a circadian rhythm. An example is the heartbeat.

**Winter solstice.** The shortest day of the year in the northern hemisphere, when the sun is at its most southerly point in the sky – 21 December.

**Zeitgeber.** From the German for 'time-giver', an entrainment signal.

**Zugunruhe.** From the German, meaning 'migratory restlessness'; surplus activity (usually at night) shown in captive birds at the time that they would normally migrate in the wild.

# APPENDIX I

While studying the cocklebur in the late 1930s, Karl Hamner and James Bonner devised a protocol of light and dark cycles to study the photoperiod effect. Hamner later refined the experimental design with the help of K. K. Nanda, an Indian biologist who spent some time in the USA. The Nanda–Hamner protocol has become one of the mainstay techniques in research into circadian and seasonal timing. In the Nanda–Hamner or 'resonance' protocol, a short main light period is coupled with a dark period of an increasing number of hours. The initial main light period is usually 6–12 hours long, and the variable dark period can be extended to give an overall period length ($T$) of about 18–72 hours. When exposed to such cycles, photoperiodic induction occurs in plants, birds and mammals when $T$ is close to 24, 48 or 72 hours; that is when the period is close to 24 hours or multiples of 24 hours. Periods such as 36 or 60 hours are non-inductive. This finding is consistent with the involvement of an endogenous 24-hour circadian pacemaker rather than an hourglass timer, which would not be expected to show a 24-hour gating of photoperiodic induction.

Night interruption experiments and skeleton photoperiods are a development of this protocol. In these experiments the main light period is usually 6–12 hours long and the extended night is systematically interrupted by a short, supplementary or scanning light pulse. The total period is at least 24 hours, but it can be much longer. For example, in the Bünsow protocol the dark period is extended to 72 hours or more. The scanning pulses in Bünsow experiments, or the main photophases in Nanda–Hamner experiments,

(A) The Nanda-Hamner or "resonance" protocol

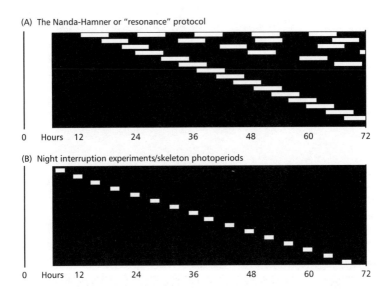

0    Hours   12           24           36           48           60           72

(B) Night interruption experiments/skeleton photoperiods

0    Hours   12           24           36           48           60           72

**Figure 1** (A) The Nanda–Hamner protocol uses a 6–12-hour main photoperiod coupled to an increasing number of dark hours. (B) The night interruption/skeleton photoperiod protocol often uses a 6–12-hour main photoperiod, and the dark component is systematically interrupted by a light pulse about 1 hour long. Each line represents a different photoperiod, and groups of animals or plants are exposed to the photoperiod for an extended period. In both the Nanda–Hamner and skeleton photoperiod protocols, if photoperiodic induction occurs at near 24 hours or multiples of 24 hours, then an underlying circadian timer is implicated. No such gated rhythmicity would be predicted by an hourglass timer.

'come on' at sequentially later phases, thereby probing the extended dark period for a light-sensitive phase that might (or might not) occur and recur as a free-running circadian oscillation. If photoperiodic induction occurs at 24 hours or multiples of 24 hours, a circadian timer rather than an hourglass timer is implicated.

# REFERENCES

Abel, E. L. & Kruger, M. L. (2004) Relation of handedness with season of birth of professional baseball players revisited. *Percept Mot Skills*, **98**, 44–46.

Aitken, A. (2004) *Sardine Run – The Greatest Shoal on Earth*. Ocean Planet, Pretoria.

Ajdacic-Gross, V., Bopp, M., Sansossio, R., Lauber, C., Gostynski, M., Eich, D., Gutzwiller, F. & Rossler, W. (2005) Diversity and change in suicide seasonality over 125 years. *J Epidemiol Community Health*, **59**, 967–72.

Åkesson, S. & Hedenström, A. (2007) How migrants get there: migratory performance and orientation. *BioScience*, **57**, 123–33.

Anderson, H. R., Bailey, P. A. & Bland, J. M. (1981) The effect of birth month on asthma, eczema, hayfever, respiratory symptoms, lung function, and hospital admissions for asthma. *Int J Epidemiol*, **10**, 45–51.

Andrade-Narvaez, F. J., Canto Lara, S. B., Van Wynsberghe, N. R., Rebollar-Tellez, E. A., Vargas-Gonzalez, A. & Albertos-Alpuche, N. E. (2003) Seasonal transmission of *Leishmania* (*Leishmania*) *mexicana* in the state of Campeche, Yucatan Peninsula, Mexico. *Mem Inst Oswaldo Cruz*, **98**, 995–98.

Arehart-Treichel, J. (2003) Long, sunny days linked to suicide incidence. *Psychiatric News*, **38**, 26.

Arthur, W. (2004) *Biased Embryos and Evolution*. Cambridge University Press, Cambridge.

Austin, C. R. & Short, R. V. (1985) *Reproduction in Mammals*, Book 4 (*Reproductive Fitness*). Cambridge University Press, Cambridge.

Aveni, A. (2000) *Empires of Time: Calendars, Clocks, and Cultures*. Tauris Parke, London.

Avison, J. (2000) *Physics for CXC*. Nelson Thornes, Cheltenham.

Axelsson, J., Stefansson, J. G., Magnusson, A., Sigvaldason, H. & Karlsson, M. M. (2002) Seasonal affective disorders: relevance of Icelandic and Icelandic-Canadian evidence to etiologic hypotheses. *Can J Psychiatry*, **47**, 153–58.

Azevedo, E., Ribeiro, J. A., Lopes, F., Martins, R. & Barros, H. (1995) Cold: a risk factor for stroke? *J Neurol*, **242**, 217–21.

Bailey, R. C., Jenike, M. R., Ellison, P. T., Bentley, G. R., Harrigan, A. M. & Peacock, N. R. (1992) The ecology of birth seasonality among agriculturalists in central Africa. *J Biosoc Sci*, **24**, 393–412.

Banerjee, A., Kalghatgi, A. T., Saiprasad, G. S., Nagendra, A., Panda, B. N., Dham, S. K., Mahen, A., Menon, K. D. & Khan, M. A. (2005) Outbreak of pneumococcal pneumonia among military recruits. *Med J Armed Forces India*, **61**, 16–21.

Banks, J. (2006) *Season Creep: How Global Warming is Already Affecting the World Around Us*. Clear the Air, Washington DC.

Barbraud, C. & Weimerskirch, H. (2001) Emperor penguins and climate change. *Nature*, **411**, 183–86.

Barbraud, C. & Weimerskirch, H. (2006) Antarctic birds breed later in response to climate change. *Proc Natl Acad Sci USA*, **103**, 6248–51.

Barrett, P., Ebling, F. J., Schuhler, S., Wilson, D., Ross, A. W., Warner, A., Jethwa, P., Boelen, A., Visser, T. J., Ozanne, D. M. *et al.* (2007) Hypothalamic thyroid hormone catabolism acts as a gatekeeper for the seasonal control of body weight and reproduction. *Endocrinology*, **148**, 3608–17.

Basso, O., Olsen, J., Bisanti, L., Juul, S. & Boldsen, J. (1995) Are seasonal preferences in pregnancy planning a source of bias in studies of seasonal variation in reproductive outcomes? The European Study Group on Infertility and Subfecundity. *Epidemiology*, **6**, 520–24.

Battle, Y. L., Martin, B. C., Dorfman, J. H. & Miller, L. S. (1999) Seasonality and infectious disease in schizophrenia: the birth hypothesis revisited. *J Psychiatr Res*, **33**, 501–09.

Beeby, A. & Brennan, A.-M. (2003) *First Ecology: Ecological Principles and Environmental Issues*, 2nd edn. Oxford University Press, Oxford.

Berteaux, D., Réale, D., McAdam, A. G. & Boutin, S. (2004) Keeping pace with fast climate change: can Arctic life count on evolution? *Integrat Comp Biol*, **44**, 140–51.

Berthold, P. & Querner, U. (1981) Genetic basis of migratory behavior in European Warblers. *Science*, **212**, 77–79.

Berthold, P., Gwinner, E. & Sonnenschein, E., eds. (2003) *Avian Migration*. Springer, Berlin.

Biller, J., Jones, M. P., Bruno, A., Adams, H. P. Jr & Banwart, K. (1988) Seasonal variation of stroke – does it exist? *Neuroepidemiology*, **7**, 89–98.

Billings, H. J., Viguie, C., Karsch, F. J., Goodman, R. L., Connors, J. M. & Anderson, G. M. (2002) Temporal requirements of thyroid hormones for seasonal changes in LH secretion. *Endocrinology*, **143**, 2618–25.

Bingman, V. P., Hough, G. E. II, Kahn, M. C. & Siegel, J. J. (2003) The homing pigeon hippocampus and space: in search of adaptive specialization. *Brain Behav Evol*, **62**, 117–27.

Blom, L., Dahlquist, G., Nystrom, L., Sandstrom, A. & Wall, S. (1989) The Swedish childhood diabetes study – social and perinatal determinants for diabetes in childhood. *Diabetologia*, **32**, 7–13.

Bonner, J. T. (1974) *On Development: The Biology of Form*. Harvard University Press, Cambridge, Massachusetts.

Borthwick, H. A., Hendricks, S. B., Parker, M. W., Toole, E. H. & Toole, V. K. (1952) A reversible photoreaction controlling seed germination. *Proc Natl Acad Sci USA*, **38**, 662–66.

Both, C. & Visser, M. E. (2001) Adjustment to climate change is constrained by arrival date in a long-distance migrant bird. *Nature*, **411**, 296–98.

Both, C., Artemyev, A. V., Blaauw, B., Cowie, R. J., Dekhuijzen, A. J., Eeva, T., Enemar, A., Gustafsson, L., Ivankina, E. V., Jarvinen, A. *et al.*

(2004) Large-scale geographical variation confirms that climate change causes birds to lay earlier. *Proc Biol Sci*, **271**, 1657–62.

Bowen, M. F., Saunders, D. S., Bollenbacher, W. E. & Gilbert, L. I. (1984) In vitro reprogramming of the photoperiodic clock in an insect brain–retrocerebral complex. *Proc Natl Acad Sci USA*, **81**, 5881–84.

Bradshaw, W. E. & Holzapfel, C. M. (2001) Genetic shift in photoperiodic response correlated with global warming. *Proc Natl Acad Sci USA*, **98**, 14509–11.

Bradshaw, W. E. & Holzapfel, C. M. (2006) Climate change. Evolutionary response to rapid climate change. *Science*, **312**, 1477–78.

Brady, J. (1979) *Biological clocks. Institute of Biology Studies in Biology*, vol. 104. Edward Arnold, London.

Brain, C. K. & Sillen, A. (1988) Evidence from Swartkrans cave for the earliest use of fire. *Nature*, **336**, 464–66.

Brandstaetter, R. & Krebs, J. (2004) Obituary: Eberhard Gwinner (1938–2004). *Nature*, **432**, 687.

Bridgman, H. A. & Oliver, J. E. (2006) *The Global Climate System: Patterns, Processes, and Teleconnections*. Cambridge University Press, Cambridge.

Briggs, W. R. & Olney, M. A. (2001) Photoreceptors in plant photomorphogenesis to date: five phytochromes, two cryptochromes, one phototropin, and one superchrome. *Plant Physiol*, **125**, 85–88.

Bronson, F. H. (2004) Are humans seasonally photoperiodic? *J Biol Rhythms*, **19**, 180–92.

Brower, L. (1996) Monarch butterfly orientation: missing pieces of a magnificent puzzle. *J Exp Biol*, **199**, 93–103.

Brown, A. S., Begg, M. D., Gravenstein, S., Schaefer, C. A., Wyatt, R. J., Bresnahan, M., Babulas, V. P. & Susser, E. S. (2004) Serologic evidence of prenatal influenza in the etiology of schizophrenia. *Arch Gen Psychiatry*, **61**, 774–80.

Bünning, E. (1973) *The Physiological Clock – Circadian rhythms and biological chronometry*, 3rd edn. Springer-Verlag, New York.

Cajochen, C. (2007) Alerting effects of light. *Sleep Med Rev*, **11**, 453–64.

Campbell, H. A., Fraser, K. P. P., Bishop, C. M., Peck, L. S. & Egginton, S. (2008) Hibernation in an Antarctic fish: on ice for winter. *PLoS ONE*, **3**(3), e1743. doi:1710.1371/journal.pone.0001743.

Cane, M. A. & Molnar, P. (2001) Closing of the Indonesian seaway as a precursor to east African aridification around 3–4 million years ago. *Nature*, **411**, 157–62.

Case, M. (2008) The essential joy of life is seeing season's unfolding. *The Observer*, London, 10 July 2005.

Castrogiovanni, P., Iapichino, S., Pacchierotti, C. & Pieraccini, F. (1998) Season of birth in psychiatry. A review. *Neuropsychobiology*, **37**, 175–81.

Cauvin, J. F. (1815) *Des Bienfaits de l'Insolation*. Paris.

Cavallaro, J. J. & Monto, A. S. (1970) Community-wide outbreak of infection with a 229E-like coronavirus in Tecumseh, Michigan. *J Infect Dis*, **122**, 272–79.

CDC (2005) Homicide and suicide rates – National Violent Death Reporting System, six states, 2003. *Morbid Mortal Wkly Rep*, **54**, 377–80.

Chariyalertsak, S., Sirisanthana, T., Supparatpinyo, K. & Nelson, K. E. (1996) Seasonal variation of disseminated *Penicillium marneffei* infections in northern Thailand: a clue to the reservoir? *J Infect Dis*, **173**, 1490–93.

Charmantier, A., McCleery, R. H., Cole, L. R., Perrins, C., Kruuk, L. E. & Sheldon, B. C. (2008) Adaptive phenotypic plasticity in response to climate change in a wild bird population. *Science*, **320**, 800–03.

Chowers, Y., Odes, S., Bujanover, Y., Eliakim, R., Bar Meir, S. & Avidan, B. (2004) The month of birth is linked to the risk of Crohn's disease in the Israeli population. *Am J Gastroenterol*, **99**, 1974–76.

Clutton-Brock, T. H., ed. (1998) *Reproductive Success: Studies of Individual Variation in Contrasting Breeding Systems*. University of Chicago Press, Chicago.

Clutton-Brock, T. H., Brotherton, P. N., Russell, A. F., O'Riain, M. J., Gaynor, D., Kansky, R., Griffin, A., Manser, M., Sharpe, L., McIlrath, G. M. *et al.* (2001) Cooperation, control, and concession in meerkat groups. *Science*, **291**, 478–81.

Cohen, P., Wax, Y. & Modan, B. (1983) Seasonality in the occurrence of breast cancer. *Cancer Res*, **43**, 892–96.

Cohen, S. (1995) Psychological stress and susceptibility to upper respiratory infections. *Am J Respir Crit Care Med*, **152**, S53–58.

Collins, K. J. (1993) Seasonal mortality in the elderly. In Ulijaszek, S. J. & Strickland, S. S. (eds) *Seasonality and Human Ecology (Society for the Study of Human Biology Symposium Series)*. Cambridge University Press, Cambridge, pp. 135–48.

Condon, R. G. & Scaglion, R. (1982) The ecology of human birth seasonality. *Hum Ecol*, **10**, 495–511.

Corbesier, L. & Coupland, G. (2006) The quest for florigen: a review of recent progress. *J Exp Bot*, **57**, 3395–403.

Dajani, Y. F., Masoud, A. A. & Barakat, H. F. (1989) Epidemiology and diagnosis of human brucellosis in Jordan. *J Trop Med Hyg*, **92**, 209–14.

Danks, H. V. (2005) Key themes in the study of seasonal adaptations in insects II. Life-cycle patterns. *Appl Entomol Zool*, **41**, 1–13.

Dauvilliers, Y., Carlander, B., Molinari, N., Desautels, A., Okun, M., Tafti, M., Montplaisir, J., Mignot, E. & Billiard, M. (2003) Month of birth as a risk factor for narcolepsy. *Sleep*, **26**, 663–65.

Davenport, J. (1992) *Animal Life at Low Temperature*. Chapman & Hall, London.

Dawson, A., Goldsmith, A. R., Nicholls, T. J. & Follett, B. K. (1986) Endocrine changes associated with the termination of photorefractoriness by short daylengths and thyroidectomy in starlings (*Sturnus vulgaris*). *J Endocrinol*, **110**, 73–79.

Dawson, A., King, V. M., Bentley, G. E. & Ball, G. F. (2001) Photoperiodic control of seasonality in birds. *J Biol Rhythms*, **16**, 365–80.

de Grazia, S. (1962) *On Time, Work, and Leisure*. Doubleday, Garden City, New York.

de Looper, M. (2002) *Seasonality of Death*. Australian Institute of Health and Welfare (AIHW), Canberra.

de Mairan, J. J. O. (1729) Observation botanique. In *Histoire de l'Académie Royale des Sciences*, pp. 35–36.

de Quattro, J. (1991) *Tripping the Light Switch Fantastic*. Agricultural Research, US Department of Agriculture.

de Waal, F. (2005) All in the family. *Science & Spirit* (September/ October). <http://www.science-spirit.org/article_detail. php?article_id=544>

Deinlein, M. (1997) *Have Wings, Will Travel: Avian Adaptations to Migration.* Migratory Bird Center (Smithsonian National Zoological Park), Washington DC, Fact sheet 4.

Delhoume, L. (1939) *De Claude Bernard à d'Arsonval.* J. B. Baillière, Paris.

Dickinson, E. (1999) *The Poems of Emily Dickinson: Reading Edition.* The Belknap Press of Harvard University Press, Cambridge, Massachusetts.

Dingle, H. & Drake, V. A. (2007) What is migration? *BioScience*, **57**, 113–21.

Dittman, A. & Quinn, T. (1996) Homing in Pacific salmon: mechanisms and ecological basis. *J Exp Biol*, **199**, 83–91.

Doblhammer, G. (1999) Longevity and month of birth: evidence from Austria and Denmark. *Demogr Res*, **1**, 1–22.

Doblhammer, G. & Vaupel, J. W. (2001) Lifespan depends on month of birth. *Proc Natl Acad Sci USA*, **98**, 2934–39.

Doblhammer, G., Rodgers, J. L. & Rau, R. (2000) Seasonality of birth in nineteenth- and twentieth-century Austria. *Soc Biol*, **47**, 201–17.

Douglas, A. S., Russell, D. & Allan, T. M. (1990) Seasonal, regional and secular variations of cardiovascular and cerebrovascular mortality in New Zealand. *Aust N Z J Med*, **20**, 669–76.

Dowell, S. F. (2001) Seasonal variation in host susceptibility and cycles of certain infectious diseases. *Emerg Infect Dis*, **7**, 369–74.

Dunnigan, M. G., Harland, W. A. & Fyfe, T. (1970) Seasonal incidence and mortality of ischaemic heart-disease. *Lancet*, **2**, 793–97.

Durkheim, E. (1897) *Suicide* (reprint edn, 1997). The Free Press, Glencoe, Illinois.

Dusheck, J. (2002) It's the ecology, stupid! *Nature*, **418**, 578–79.

Eagles, J. M. (2004) Light therapy and the management of winter depression. *Adv Psychiat Treat*, **10**, 233–40.

Eccles, R. (2002) An explanation for the seasonality of acute upper respiratory tract viral infections. *Acta Otolaryngol*, **122**, 183–91.

Eisenberg, D. T., Campbell, B., Mackillop, J., Lum, J. K. & Wilson, D. S. (2007) Season of birth and dopamine receptor gene associations with impulsivity, sensation seeking and reproductive behaviors. *PLoS ONE*, 2, e1216.

Ellison, P. T. (2003) Energetics and reproductive effort. *Am J Hum Biol*, 15, 342–51.

Emlen, S. T. (1970) Celestial rotation: its importance in the development of migratory orientation. *Science*, 170, 1198–201.

Emlen, S. T. (1975) The stellar-orientation system of a migratory bird. *Sci Am*, 233(2), 102–11.

Eskola, J., Takala, A. K., Kela, E., Pekkanen, E., Kalliokoski, R. & Leinonen, M. (1992) Epidemiology of invasive pneumococcal infections in children in Finland. *J Am Med Assoc*, 268, 3323–27.

Esquirol, J. E. (1845) *Mental Maladies, a Treatise on Insanity*. Lea & Blanchard, Philadelphia.

European Union (2001) *Eurobarometer 55.2: Europeans, Science and Technology*. EU, December.

Farner, D. S. (1950) The annual stimulus for migration. *Condor*, 52, 104–22.

Fielder, J. L. (2001) *Heliotherapy: The Principles & Practice of Sunbathing*. Academy of Natural Living, Cairns, Queensland.

Fisman, D. N., Lim, S., Wellenius, G. A., Johnson, C., Britz, P., Gaskins, M., Maher, J., Mittleman, M. A., Spain, C. V., Haas, C. N. & Newbern, C. (2005) It's not the heat, it's the humidity: wet weather increases legionellosis risk in the greater Philadelphia metropolitan area. *J Infect Dis*, 192, 2066–73.

Fitter, A. H. & Fitter, R. S. R. (2002) Rapid changes in flowering time in British plants. *Science*, 296, 1689–91.

Fleissner, G., Stahl, B., Thalau, P., Falkenberg, G. & Fleissner, G. (2007) A novel concept of Fe-mineral-based magnetoreception: histological and physicochemical data from the upper beak of homing pigeons. *Naturwissenschaften*, 94, 631–42.

Foley, R. A. (1993) The influence of seasonality on hominid evolution. In Ulijaszek, S. J. & Strickland, S. S. (eds) *Seasonality and Human Ecology (35th Symposium Volume of the Society for the Study of Human Biology)*. Cambridge University Press, Cambridge, pp. 17–38.

Follett, B. K. & Sharp, P. J. (1969) Circadian rhythmicity in photoperiodically induced gonadotrophin release and gonadal growth in the quail. *Nature*, **223**, 968–71.

Foster, R. G. (1998) Photoentrainment in the vertebrates: a comparative analysis. In Lumsden, P. J. & Millar, A. J. (eds) *Biological Rhythms and Photoperiodism In Plants*. BIOS, Oxford, pp. 135–49.

Foster, R. G. & Hankins, M. W. (2007) Circadian vision. *Curr Biol*, **17**, R746–51.

Foster, R. G. & Kreitzman, L. (2004) *Rhythms of Life: The Biological Clocks that Control the Daily Lives of Every Living Thing*. Profile Books, London.

Foster, R. G. & Roenneberg, T. (2008) Human responses to the geophysical daily, annual and lunar cycles. *Curr Biol*, **18**(17), R784–94.

Foster, R. G., Follett, B. K. & Lythgoe, J. N. (1985) Rhodopsin-like sensitivity of extra-retinal photoreceptors mediating the photoperiodic response in quail. *Nature*, **313**, 50–52.

Foster, R. G., Peirson, S. & Whitmore, D. (2004) Chronobiology. In Meyers, R. A. (ed.) *Encyclopedia of Molecular Cell Biology and Molecular Medicine*. Wiley-VCH, Weinheim, pp. 413–81.

Fraser, J. T. (1987) *Time, the Familiar Stranger*. University of Massachusetts Press, Amherst.

Froy, O., Gotter, A. L., Casselman, A. L. & Reppert, S. M. (2003) Illuminating the circadian clock in monarch butterfly migration. *Science*, **300**, 1303–05.

Galdikas, B. M. F. (1988) Orangutan diet, range, and activity at Tanjung Puting, Central Borneo. *Int J Primatol*, **9**, 1–35.

Galea, M. H. & Blamey, R. W. (1991) Season of initial detection in breast cancer. *Br J Cancer*, **63**, 157.

Garner, D. M. (1997) The effects of starvation on behaviour: implications for eating disorders. In Garner, D. M. & Garfinkel, P. E. (eds) *Handbook of Treatment for Eating Disorders*. Guilford Press, New York, pp. 145–77.

Garner, W. W. & Allard, H. A. (1920) Effect of relative length of day and night and other factors of the environment on growth and reproduction in plants. *J Agric Res*, **18**, 553–606.

Glimaker, M., Samuelson, A., Magnius, L., Ehrnst, A., Olcen, P. & Forsgren, M. (1992) Early diagnosis of enteroviral meningitis by detection of specific IgM antibodies with a solid-phase reverse immunosorbent test (SPRIST) and mu-capture EIA. *J Med Virol*, **36**, 193–201.

Gluckman, P. & Hanson, M. (2006) *Why Our World No Longer Fits Our Bodies*. Oxford University Press, Oxford.

Golden, R. N., Gaynes, B. N., Ekstrom, R. D., Hamer, R. M., Jacobsen, F. M., Suppes, T., Wisner, K. L. & Nemeroff, C. B. (2005) The efficacy of light therapy in the treatment of mood disorders: a review and meta-analysis of the evidence. *Am J Psychiatry*, **162**, 656–62.

Goldman, B. D., Darrow, J. M. & Yogev, L. (1984) Effects of timed melatonin infusions on reproductive development in the Djungarian hamster (*Phodopus sungorus*). *Endocrinology*, **114**, 2074–83.

Goodall, J. (1986) *The Chimpanzees of Gombe: Patterns of Behavior*. Harvard University Press, Boston.

Goodwin, J. (2007) A deadly harvest: the effects of cold on older people in the UK. *Br J Community Nurs*, **12**, 23–26.

Goren-Inbar, N., Alperson, N., Kislev, M. E., Simchoni, O., Melamed, Y., Ben-Nun, A. & Werker, E. (2004) Evidence of hominin control of fire at Gesher Benot Ya'aqov, Israel. *Science*, **304**, 725–27.

Gowlett, J. A. J., Harris, J. W. K., Walton, D. & Wood, B. A. (1981) Early archaeological sites, hominid remains and traces of fire from Chesowanja, Kenya. *Nature*, **294**, 125–29.

Griffin, D. R. (1964) *Bird Migration*. The Natural History Press, New York.

Grossman, D. (2004) Spring forward. *Sci Am* **290**(1), 84–91.

Gullan, P. J. & Cranston, P. (2004) *The Insects: An Outline of Entomology*. Blackwell Publishing, Oxford.

Gwinner, E. (1967) Circannuale Periodik der Mauser und der Zugunruhe bei einem Vogel. *Naturwissenschaften*, **54**, 447.

Gwinner, E. (1986) *Circannual Rhythms: Endogenous Annual Clocks in the Organisation of Seasonal Processes*. Springer-Verlag, Berlin.

Gwinner, E. (1989) Photoperiod as a modifying and limiting factor in the expression of avian circannual rhythms. *J Biol Rhythms*, **4**, 237–50.

Gwinner, E. (1996a) Circadian and circannual programmes in avian migration. *J Exp Biol*, **199**, 39–48.

Gwinner, E. (1996b) Circannual clocks in avian reproduction and migration. *Ibis*, **138**, 47–63.

Gwinner, E. (2003) Circannual rhythms in birds. *Curr Opin Neurobiol*, **13**, 770–78.

Gwinner, E. & Wiltschko, R. (1980) Circannual changes in migratory orientation of the Garden Warbler, *Sylvia borin*. *Behav Ecol Sociobiol*, **7**, 73–78.

Hafner, H., Haas, S., Pfeifer-Kurda, M., Eichhorn, S. & Michitsuji, S. (1987) Abnormal seasonality of schizophrenic births. A specific finding? *Eur Arch Psychiatry Neurol Sci*, **236**, 333–42.

Hall, C. B., Walsh, E. E., Long, C. E. & Schnabel, K. C. (1991) Immunity to and frequency of reinfection with respiratory syncytial virus. *J Infect Dis*, **163**, 693–98.

Hambre, D. & Beem, M. (1972) Virologic studies of acute respiratory disease in young adults. V. Coronavirus 229E infections during six years of surveillance. *Am J Epidemiol*, **96**, 94–106.

Hamner, K. C. & Bonner, J. (1938) Photoperiodism in regulation to hormones as factors in floral initiation and development. *Bot Gaz*, **100**, 388–431.

Hangarter, R. P. & Gest, H. (2004) Pictorial demonstrations of photosynthesis. *Photosynth Res*, **80**, 421–25.

Hardie, J. & Vaz Nunes, M. (2001) Aphid photoperiodic clocks. *J Insect Physiol*, **47**, 821–32.

Hare, E. & Moran, P. (1981) A relation between seasonal temperature and the birth rate of schizophrenic patients. *Acta Psychiatr Scand*, **63**, 396–405.

Hare, E., Price, J. & Slater, E. (1974) Mental disorder and season of birth: a national sample compared with the general population. *Br J Psychiatry*, **124**, 81–86.

Harrison, S. (2004) Emotional climates: ritual, seasonality and affective disorders. *J R Anthropol Inst*, **10**, 583–602.

Hattar, S., Lucas, R. J., Mrosovsky, N., Thompson, S., Douglas, R. H., Hankins, M. W., Lem, J., Biel, M., Hofmann, F., Foster, R. G. & Yau,

K. W. (2003) Melanopsin and rod–cone photoreceptive systems account for all major accessory visual functions in mice. *Nature*, **424**, 76–81.

Hau, M., Wikelski, M. & Wingfield, J. C. (1998) A neotropical forest bird can measure the slight changes in tropical photoperiod. *Proc R Soc B*, **265**, 89–95.

Haus, E., Halberg, F., Kuhl, J. F. & Lakatua, D. J. (1974) Chronopharmacology in animals. *Chronobiologia*, **1** (Suppl 1), 122–56.

Hayes, C. E., Cantorna, M. T. & DeLuca, H. F. (1997) Vitamin D and multiple sclerosis. *Proc Soc Exp Biol Med*, **216**, 21–27.

Heerlein, A., Valeria, C. & Medina, B. (2006) Seasonal variation in suicidal deaths in Chile: its relationship to latitude. *Psychopathology*, **39**, 75–79.

Helbig, A. J., Berthold, P. & Wiltschko, R. (1989) Migratory orientation of blackcaps (*Sylvia atricapilla*): population-specific shifts of direction during the autumn. *Ethology*, **82**, 307–15.

Heller, H. C. & Ruby, N. F. (2004) Sleep and circadian rhythms in mammalian torpor. *Annu Rev Physiol*, **66**, 275–89.

Hendley, J. O., Fishburne, H. B. & Gwaltney, J. M. Jr (1972) Coronavirus infections in working adults. Eight-year study with 229 E and OC 43. *Am Rev Respir Dis*, **105**, 805–11.

Hiebert, S. M., Thomas, E. M., Lee, T. M., Pelz, K. M., Yellon, S. M. & Zucker, I. (2000) Photic entrainment of circannual rhythms in golden-mantled ground squirrels: role of the pineal gland. *J Biol Rhythms*, **15**, 126–34.

Hippocrates (400 BC) *Nature of Man*, vol. IV (translated by W. H. S. Jones, 1931) Loeb Classical Library.

Hoffman, K. (1954) Versuche zu der im Richtungsfinden der Vögel enthaltenen Zeitschätzung. *Z Tierpsychol*, **11**, 453–75.

Hostmark, J. G., Laerum, O. D. & Farsund, T. (1984) Seasonal variations of symptoms and occurrence of human bladder carcinomas. *Scand J Urol Nephrol*, **18**, 107–11.

Hrushesky, W. J. (1991) The multifrequency (circadian, fertility cycle, and season) balance between host and cancer. *Ann N Y Acad Sci*, **618**, 228–56.

Huber, S., Fieder, M., Wallner, B., Iber, K. & Moser, G. (2004) Effects of season of birth on reproduction in contemporary humans: brief communication. *Hum Reprod*, **19**, 445–47.

Hughes, L. (2000) Biological consequences of global warming: is the signal already apparent? *Trends Ecol Evol*, **15**, 56–61.

Hviid, L. (1998) Clinical disease, immunity and protection against *Plasmodium falciparum* malaria in populations living in endemic areas. *Expert Rev Mol Med*, **1998**, 1–10.

Inman, M. (2008) Great tits enjoying the warmer weather – so far. *New Scient.*, 8 May.

Irving, L. (1966) Adaptations to cold. *Sci Am*, **214**(1), 94–101.

Jordanes (AD 551) *De origine actibusque Getarum* [*The Origin and Deeds of the Goths*]. The Project Gutenberg EBook. <www.gutenberg.org/etext/14809>

Kalkstein, L. S. (2000) Biometeorology – looking at the links between weather, climate and health. *Biometeorol Bull*, **5**, 9–18.

Kalm, L. M. & Semba, R. D. (2005) They starved so that others be better fed: remembering Ancel Keys and the Minnesota experiment. *J Nutr*, **135**, 1347–52.

Kapikian, A. Z., Kim, H. W., Wyatt, R. G., Cline, W. L., Arrobio, J. O., Brandt, C. D., Rodriguez, W. J., Sack, D. A., Chanock, R. M. & Parrott, R. H. (1976) Human reovirus-like agent as the major pathogen associated with 'winter' gastroenteritis in hospitalized infants and young children. *N Engl J Med*, **294**, 965–72.

Karimi, M. & Yarmohammadi, H. (2003) Seasonal variations in the onset of childhood leukemia/lymphoma: April 1996 to March 2000, Shiraz, Iran. *Hematol Oncol*, **21**, 51–55.

Keatinge, W. R., Coleshaw, S. R. & Holmes, J. (1989) Changes in seasonal mortalities with improvement in home heating in England and Wales from 1964 to 1984. *Int J Biometeorol*, **33**, 71–76.

Kim, C. D., Lesage, A. D., Seguin, M., Chawky, N., Vanier, C., Lipp, O. & Turecki, G. (2004) Seasonal differences in psychopathology of male suicide completers. *Compr Psychiatry*, **45**, 333–39.

Kim, J., Kim, Y., Yeom, M., Kim, J.-H. & Nam, H. G. (2008) FIONA1 is essential for regulating period length in the *Arabidopsis* circadian clock. *Plant Cell*, **20**, 307–19.

Kingsley, C. (2004) *Saint's Tragedy*. Kessinger Publishing Co., Whitefish, Montana.

Klein, D. C., Smoot, R., Weller, J. L., Higa, S., Markey, S. P., Creed, G. J. & Jacobowitz, D. M. (1983) Lesions of the paraventricular nucleus area of the hypothalamus disrupt the suprachiasmatic leads to spinal cord circuit in the melatonin rhythm generating system. *Brain Res Bull*, **10**, 647–52.

Kloner, R. A. (2004) The 'Merry Christmas Coronary' and 'Happy New Year Heart Attack' phenomenon. *Circulation*, **110**, 3744–45.

Kolbert, E. (2005) The Climate of Man. In *New Yorker*, April 2005.

Kondo, N., Sekijima, T., Kondo, J., Takamatsu, N., Tohya, K. & Ohtsu, T. (2006) Circannual control of hibernation by HP complex in the brain. *Cell*, **125**, 161–172.

Koorengevel, K. M. (2001) *On the Chronobiology of Seasonal Affective Disorder*. Thesis, University of Groningen, Groningen.

Krakauer, D. (2004) What gets you going first thing? In *Times Higher Education* (London), 13 August 2004.

Kramer, G. (1952) Experiments on bird orientation. *Ibis*, **94**, 265–85.

Kramer, T. R., Moore, R. J., Shippee, R. L., Friedl, K. E., Martinez-Lopez, L., Chan, M. M. & Askew, E. W. (1997) Effects of food restriction in military training on T-lymphocyte responses. *Int J Sports Med*, **18** (Suppl 1), S84–90.

Kristoffersen, S. & Hartveit, F. (2000) Is a woman's date of birth related to her risk of developing breast cancer? *Oncol Rep*, **7**, 245–47.

Krug, E. G., Dahlberg, L. L., Mercy, J. A., Zwi, A. B. & Lozano, R. (2002) World report on violence and health. World Health Organization, Geneva.

Lacoste, V. & Wirz-Justice, A. (1989) Seasonal variation in normal subjects: an update of variables current in depression research. In Rosenthal, N. E. & Blehar, M. (eds) *Seasonal Affective Disorders and Phototherapy*. Guilford Press, New York, pp. 167–229.

Lam, D. A. & Miron, J. A. (1994) Global patterns of seasonal variation in human fertility. *Ann N Y Acad Sci*, **709**, 9–28.

Lam, R. W. & Levitan, R. D. (2000) Pathophysiology of seasonal affective disorder: a review. *J Psychiatry Neurosci*, **25**, 469–80.

Lam, R. W., Tam, E. M., Shiah, I. S., Yatham, L. N. & Zis, A. P. (2000) Effects of light therapy on suicidal ideation in patients with winter depression. *J Clin Psychiatry*, **61**, 30–32.

Lam, R. W., Levitt, A. J., Levitan, R. D., Enns, M. W., Morehouse, R., Michalak, E. E. & Tam, E. M. (2006) The Can-SAD study: a randomized controlled trial of the effectiveness of light therapy and fluoxetine in patients with winter seasonal affective disorder. *Am J Psychiatry*, **163**, 805–12.

Lambert, G. W., Reid, C., Kaye, D. M., Jennings, G. L. & Esler, M. D. (2002) Effect of sunlight and season on serotonin turnover in the brain. *Lancet*, **360**, 1840–42.

Langagergaard, V., Norgard, B., Mellemkjaer, L., Pedersen, L., Rothman, K. J. & Sorensen, H. T. (2003) Seasonal variation in month of birth and diagnosis in children and adolescents with Hodgkin disease and non-Hodgkin lymphoma. *J Pediatr Hematol Oncol*, **25**, 534–38.

Lantz, T. C. & Turner, N. J. (2003) Traditional phenological knowledge of aboriginal peoples in British Columbia. *J Ethnobiol*, **23**, 263–86.

Lechan, R. M. & Fekete, C. (2005) Role of thyroid hormone deiodination in the hypothalamus. *Thyroid*, **15**, 883–97.

Lechowicz, M. J. (2001) *Phenology*. Wiley, London.

Lees, A. D. (1964) The location of the photoperiodic receptors in the aphid *Megoura viciae* Buckton. *J Exp Biol*, **41**, 119–33.

Lees, A. D. (1966) Photoperiodic timing mechanisms in insects. *Nature*, **210**, 986–89.

Leffingwell, A. (1892) *Illegitimacy and the Influence of Seasons upon Conduct: Two Studies in Demography*. Swan Sonnenschein & Co, London.

Lerchl, A., Simoni, M. & Nieschlag, E. (1993) Changes in seasonality of birth rates in Germany from 1951 to 1990. *Naturwissenschaften*, **80**, 516–18.

Lewis-Williams, J. D. (2002) *The Mind in the Cave: Consciousness and the Origins of Art*. Thames & Hudson, London.

Lewy, A. J., Wehr, T. A., Goodwin, F. K., Newsome, D. A. & Markey, S. P. (1980) Light suppresses melatonin secretion in humans. *Science*, **210**, 1267–69.

Lewy, A. J., Kern, H. A., Rosenthal, N. E. & Wehr, T. A. (1982) Bright artificial light treatment of a manic-depressive patient with a seasonal mood cycle. *Am J Psychiatry*, **139**, 1496–98.

Lewy, A. J., Lefler, B. J., Emens, J. S. & Bauer, V. K. (2006) The circadian basis of winter depression. *Proc Natl Acad Sci USA*, **103**, 7414–19.

Lincoln, G. A. (1998) Reproductive seasonality and maturation throughout the complete life-cycle in the mouflon ram (*Ovis musimon*). *Anim Reprod Sci*, **53**, 87–105.

Lincoln, G. A., Andersson, H. & Loudon, A. (2003) Clock genes in calendar cells as the basis of annual timekeeping in mammals – a unifying hypothesis. *J Endocrinol*, **179**, 1–13.

Lincoln, G. A., Johnston, J. D., Andersson, H., Wagner, G. & Hazlerigg, D. G. (2005) Photorefractoriness in mammals: dissociating a seasonal timer from the circadian-based photoperiod response. *Endocrinology*, **146**, 3782–90.

Lincoln, G. A., Clarke, I. J., Hut, R. A. & Hazlerigg, D. G. (2006) Characterizing a mammalian circannual pacemaker. *Science*, **314**, 1941–44.

Lindén, L. (2002) *Measuring Cold Hardiness in Woody Plants*. Faculty of Agriculture and Forestry, Department of Applied Biology, University of Helsinki, Helsinki.

Lofts, B. (1970) *Animal Photoperiodism*. Edward Arnold, London.

Lohmann, K. J., Lohmann, C. M. & Putman, N. F. (2007) Magnetic maps in animals: nature's GPS. *J Exp Biol*, **210**, 3697–3705.

Lowen, A. C., Mubareka, S., Steel, J. & Palese, P. (2007) Influenza virus transmission is dependent on relative humidity and temperature. *PLoS Pathog*, **3**, 1470–76.

Lumey, L. H. & Stein, A. D. (1997) *In utero* exposure to famine and subsequent fertility: the Dutch Famine Birth Cohort Study. *Am J Public Health*, **87**, 1962–66.

Lummaa, V. & Tremblay, M. (2003) Month of birth predicted reproductive success and fitness in pre-modern Canadian women. *Proc Biol Sci*, **270**, 2355–61.

Macgregor, D. J. & Lincoln, G. A. (2008) A physiological model of a circannual oscillator. *J Biol Rhythms*, **23**, 252–64.

Madden, P. A., Heath, A. C., Rosenthal, N. E. & Martin, N. G. (1996) Seasonal changes in mood and behavior. The role of genetic factors. *Arch Gen Psychiatry*, **53**, 47–55.

Marcovitch, S. (1924) The migration of the Aphididae and the appearance of the sexual forms as affected by the relative length of daily light exposure. *J Agric Res*, **27**, 513–22.

Martinet, L., Allain, D. & Weiner, C. (1984) Role of prolactin in the photoperiodic control of moulting in the mink (*Mustela vison*). *J Endocrinol*, **103**, 9–15.

Marx, H. (1946) Hypophysare Insuffizienz bei Lichtmangel. Zur Klinik des Hypophysenzwischenhirnsystems. *Klin Wschr*, **24/25**, 18–21.

Mason, B. H., Holdaway, I. M., Mullins, P. R., Kay, R. G. & Skinner, S. J. (1985) Seasonal variation in breast cancer detection: correlation with tumour progesterone receptor status. *Breast Cancer Res Treat*, **5**, 171–76.

Mathews-Amos, A. & Berntson, D. A. (1999) *Turning up the Heat: How Global Warming Threatens Life in the Sea*. Diane Publishing, Darby, Pennsylvania.

Mathias, D., Jacky, L., Bradshaw, W. E. & Holzapfel, C. M. (2007) Quantitative trait loci associated with photoperiodic response and stage of diapause in the pitcher-plant mosquito, *Wyeomyia smithii*. *Genetics*, **176**, 391–402.

Matthews, K., Shepherd, J. & Sivarajasingham, V. (2006) Violence-related injury, and the price of beer in England and Wales. *Appl Econ*, **38**, 661–70.

McGrath, J. (1999) Hypothesis: is low prenatal vitamin D a risk-modifying factor for schizophrenia? *Schizophr Res*, **40**, 173–77.

McNab, B. K. (2002) *The Physological Ecology of Vertebrates: A View from Energetics*. Cornell University Press, Cornell.

McWhirter, W. R. & Dobson, C. (1995) Childhood melanoma in Australia. *World J Surg*, **19**, 334–36.

Mech, L. D. & Boitani, L. (2003) *Wolves: Behaviour, Ecology, and Conservation*. University of Chicago Press, Chicago.

Meier, C. R., Jick, S. S., Derby, L. E., Vasilakis, C. & Jick, H. (1998) Acute respiratory-tract infections and risk of first-time acute myocardial infarction. *Lancet*, **351**, 1467–71.

Menzel, A., Sparks, T. H., Estrella, N., Koch, E., Aasa, A., Ahas, R., Alm-Kuber, K., Bissolli, P., Braslavska, O., Briede, A. *et al.* (2006) European phenological response to climate change matches the warming pattern. *Global Change Biology*, **12**, 1969–76.

Meriggiola, M. C., Noonan, E. A., Paulsen, C. A. & Bremner, W. J. (1996) Annual patterns of luteinizing hormone, follicle stimulating hormone, testosterone and inhibin in normal men. *Hum Reprod*, **11**, 248–52.

Merkel, F. W. & Wiltschko, W. (1965) Magnetismus and Richtungsfinden zugunruhiger Rotkehlchen (*Erithacus rubecula*). Vogelwarte **23**, 71–77.

Michael, T. P. & McClung, C. R. (2003) Enhancer trapping reveals widespread circadian clock transcriptional control in *Arabidopsis*. *Plant Physiol*, **132**, 629–39.

Millar, R. & Free, E. E. (2004) *Sunrays And Health: Every Day Use of Natural And Artificial Ultraviolet Light* (reprint of 1929 edition). Kessinger Publishing Co., Whitefish, Montana.

Mithen, S. (2005) *The Prehistory of the Mind: A Search for the Origins of Art, Religion and Science*. Thames & Hudson, London.

Moore, J. A. (1939) Temperature tolerance and areas of development in the eggs of Amphibia. *Ecology*, **20**, 459–78.

Moore, S. E., Cole, T. J., Collinson, A. C., Poskitt, E. M., McGregor, I. A. & Prentice, A. M. (1999) Prenatal or early postnatal events predict infectious deaths in young adulthood in rural Africa. *Int J Epidemiol*, **28**, 1088–95.

Moran, M. (2005) Light therapy kept in dark despite effectiveness. *Psychiatric News*, **40**, 29.

Morgan, P. J. & Hazlerigg, D. G. (2008) Photoperiodic signalling through the melatonin receptor turns full circle. *J Neuroendocrinol*, **20**, 820–26.

Morken, G. & Linaker, O. M. (2000) Seasonal variation of violence in Norway. *Am J Psychiatry*, **157**, 1674–78.

Mrosovsky, N. (1988) Seasonal affective disorder, hibernation, and annual cycles in animals: chipmunks in the sky. *J Biol Rhythms*, **3**, 189–207.

Mrosovsky, N. (1990) *Rheostasis: The Physiology of Change*, 1st edn. Oxford University Press, New York.

Nakao, N., Ono, H., Yamamura, T., Anraku, T., Takagi, T., Higashi, K., Yasuo, S., Katou, Y., Kageyama, S., Uno, Y. *et al.* (2008) Thyrotrophin in the pars tuberalis triggers photoperiodic response. *Nature*, **452**, 317–22.

Nathan, A. & Barbosa, V. C. (2008) V-like formations in flocks of artificial birds. *Artif Life*, **14**, 179–88.

Nelson, R. J., Demas, G. E., Klein, S. L. & Kriegsfeld, L. J. (2002) *Seasonal Patterns of Stress, Immune Function, and Disease*. Cambridge University Press, Cambridge.

Newton, I. (2008) *The Migration Ecology of Birds*. Academic Press/ Elsevier, London.

Newton-Fisher, N. E. (1999) The diet of chimpanzees in the Budongo Forest Reserve, Uganda. *Afr J Ecol*, **37**, 344–54.

Nicolas, R. (2004) Was the emergence of home bases and domestic fire a punctuated event? A review of the Middle Pleistocene record in Eurasia. *J Archaeol Asia Pacific*, **43**, 248–81.

Nisimura, T. & Numata, H. (2003) Circannual control of the life cycle in the Varied Carpet Beetle *Anthrenus verbasci*. *Funct Ecol*, **17**, 489–95.

Norris, D. R., Marra, P. P., Kyser, T. K., Sherry, T. W. & Ratcliffe, L. M. (2004) Tropical winter habitat limits reproductive success on the temperate breeding grounds in a migratory bird. *Proc R Soc B*, **271**, 59–64.

Norris, S., Rosentrater, L. & Eid, M. P. (2002) *Polar Bears at Risk*. WWF – World Wide Fund for Nature, Gland, Switzerland.

Notter, D. R. (2002) Opportunities to reduce seasonality of breeding in sheep by selection. *Sheep Goat Res J*, **17**, 20–32.

Noyes, R. (1973) Seneca on death. *J Religion Health*, **12**, 223–40.

O'Reilly, J. (1980) In Manhattan: mink is no four-letter word. In *Time Magazine*, 18 February.

Oliver, J. (2006) The myth of Thomas Szasz. *New Atlantis*, no. 13, 68–84.

Olson. J. (2005) Quoted in *Avoid Getting Stung: Summertime Mosquito Season*. Agricultural Communications, Texas A&M University, College Station.

Papadopoulos, F. C., Frangakis, C. E., Skalkidou, A., Petridou, E., Stevens, R. G. & Trichopoulos, D. (2005) Exploring lag and duration effect of sunshine in triggering suicide. *J Affect Disord*, **88**, 287–97.

Parmesan, C. & Yohe, G. (2003) A globally coherent fingerprint of climate change impacts across natural systems. *Nature*, **421**, 37–42.

Pedersen, P. A. & Weeke, E. R. (1983) Month of birth in asthma and allergic rhinitis. *Scand J Prim Health Care*, **1**, 97–101.

Pell, J. P. & Cobbe, S. M. (1999) Seasonal variations in coronary heart disease. *Q J Med*, **92**, 689–96.

Pengelley, E. T. & Fisher, K. C. (1966) Locomotor activity patterns and their relation to hibernation in the golden-mantled ground squirrel. *J Mammal*, **47**, 63–73.

Pengelley, E. T., Asmundson, S. J., Barnes, B. M. & Aloia, R. C. (1976) Relationship of light intensity and photoperiod to circannual rhythmicity in the hibernating ground squirrel, *Citellus lateralis. Comp Biochem Physiol*, **53A**, 273–77.

Petridou, E., Papadopoulos, F. C., Frangakis, C. E., Skalkidou, A. & Trichopoulos, D. (2002) A role of sunshine in the triggering of suicide. *Epidemiology*, **13**, 106–09.

Phillips, D. P., Van Voorhees, C. A. & Ruth, T. E. (1992) The birthday: lifeline or deadline? *Psychosom Med*, **54**, 532–42.

Phillips, D. P., Jarvinen, J. R., Abramson, I. S. & Phillips, R. R. (2004) Cardiac mortality is higher around Christmas and New Year's than at any other time: the holidays as a risk factor for death. *Circulation*, **110**, 3781–88.

Piersma, T. & Gill, R. E. (1998) Guts don't fly: small digestive organs in obese Bar-tailed Godwits. *Auk*, **115**, 196–203.

Pietinalho, A., Ohmichi, M., Hiraga, Y., Lofroos, A. B. & Selroos, O. (1996) The mode of presentation of sarcoidosis in Finland and Hokkaido, Japan. A comparative analysis of 571 Finnish and 686 Japanese patients. *Sarcoidosis Vasc Diffuse Lung Dis*, **13**, 159–66.

Pinel, P. (1806) *A Treatise on Insanity*. Cadell & Davies, Sheffield.

Post, E. & Forchhammer, M. C. (2008) Climate change reduces reproductive success of an Arctic herbivore through trophic mismatch. *Phil Trans R Soc B*, **363**, 2369–75.

Post, E., Pedersen, C., Wilmers, C. C. & Forchhammer, M. C. (2008) Warming, plant phenology and the spatial dimension of trophic mismatch for large herbivores. *Proc R Soc B*, **275**, 2005–13.

Pough, F. H., Hiser, J. B. & McFarland, W. N. (1996) *Vertebrate Life*, 4th edn. Prentice Hall, Upper Saddle River, New Jersey.

Pray, L. A. (2004) Epigenetics: genome, meet your environment. *Scientist*, **18**, 14.

Press, F., Siever, R., Grotzinger, J. & Jordan, T. H. (2003) *Understanding Earth*, 4th edn. W. H. Freeman & Co Ltd, London.

Ramos, A., Perez-Solis, E., Ibanez, C., Casado, R., Collada, C., Gomez, L., Aragoncillo, C. & Allona, I. (2005) Winter disruption of the circadian clock in chestnut. *Proc Natl Acad Sci USA*, **102**, 7037–42.

Rau, R. & Doblhammer, D. (2003) Seasonal mortality in Denmark: the role of sex and age. *Demogr Res*, **9**, 197–222.

Raven, P. H., Evert, R. F. & Eichhorn, S. E. (1999) *Biology of Plants*. Freeman, New York.

Reinberg, A., Schuller, E. & Delasnerie, N. (1977) Rythmes circadiens et circannuels des leucocytes, protéines totales, immunoglobulines A, G et M. *Nouv Presse Med*, **6**, 3819–23.

Reiter, R. J. (1975) Exogenous and endogenous control of the annual reproductive cycle in the male golden hamster: participation of the pineal gland. *J Exp Zool*, **191**, 111–20.

Reppert, S. M. (2006) A colorful model of the circadian clock. *Cell*, **124**, 233–36.

Reuters (2003) Experts look to Australia's Aborigines for weather help. <http://www.climateark.org/shared/reader/welcome.aspx?linkid=21301>

Ridley, M. (2004) *Nature via Nurture: Genes, Experience and What Makes Us Human*. Harper Perennial, London.

Rihmer, Z., Rutz, W., Pihlgren, H. & Pestality, P. (1998) Decreasing tendency of seasonality in suicide may indicate lowering rate of depressive suicides in the population. *Psychiatry Res*, **81**, 233–40.

Rock, D., Greenberg, D. & Hallmayer, J. (2006) Season-of-birth as a risk factor for the seasonality of suicidal behaviour. *Eur Arch Psychiatry Clin Neurosci*, **256**, 98–105.

Rodin, I. & Martin, N. (1998) Seasonal affective disorder. *Perspect Depression*, March, 6–9.

Roenneberg, T. (2004) The decline in human seasonality. *J Biol Rhythms*, **19**, 193–95; discussion 196–97.

Roenneberg, T. & Aschoff, J. (1990) Annual rhythm of human reproduction. II. Environmental correlations. *J Biol Rhythms*, **5**, 217–39.

Roenneberg, T. & Foster, R. G. (1997) Twilight times: light and the circadian system. *Photochem Photobiol*, **66**, 549–61.

Roseboom, T. J., van der Meulen, J. H., Osmond, C., Barker, D. J., Ravelli, A. C. & Bleker, O. P. (2000) Plasma lipid profiles in adults after prenatal exposure to the Dutch famine. *Am J Clin Nutr*, **72**, 1101–06.

Rosen, L. N., Targum, S. D., Terman, M., Bryant, M. J., Hoffman, H., Kasper, S. F., Hamovit, J. R., Docherty, J. P., Welch, B. & Rosenthal, N. E. (1990) Prevalence of seasonal affective disorder at four latitudes. *Psychiatry Res*, **31**, 131–44.

Rosenberg, A. M. (1988) The clinical associations of antinuclear antibodies in juvenile rheumatoid arthritis. *Clin Immunol Immunopathol*, **49**, 19–27.

Rosenthal, N. E. (2006) *Winter Blues: Everything You Need to Know to Beat Seasonal Affective Disorder*. Guilford Press, New York.

Rosenthal, N. E. & Blehar, M. C., eds. (1989) *Seasonal Affective Disorders and Phototherapy*. Guilford Press, New York.

Rothwell, P. M., Staines, A., Smail, P., Wadsworth, E. & McKinney, P. (1996) Seasonality of birth of patients with childhood diabetes in Britain. *BMJ*, **312**, 1456–57.

Rowan, W. (1925) Relation of light to bird migration and developmental changes. *Nature*, **115**, 494–95.

Rowan, W. (1929) Experiments in bird migration. I. Manipulation of the reproductive cycle. *Proc Boston Soc Nat Hist*, **19**, 151–208.

Ruby, N. F., Dark, J., Heller, H. C. & Zucker, I. (1998) Suprachiasmatic nucleus: role in circannual body mass and hibernation rhythms of ground squirrels. *Brain Res*, **782**, 63–72.

Russell, A. F., Clutton-Brock, T. H., Brotherton, P. N. M., Sharpe, L. L., Mcilrath, G. M., Dalerum, F. D., Cameron, E. Z. & Barnard, J. A. (2002) Factors affecting pup growth and survival in co-operatively breeding meerkats *Suricata suricatta. J Anim Ecol*, **71**, 700–09.

Sadovnick, A. D., Duquette, P., Herrera, B., Yee, I. M. & Ebers, G. C. (2007) A timing-of-birth effect on multiple sclerosis clinical phenotype. *Neurology*, **69**, 60–62.

Sage, L. C. (1992) *Pigment of the Imagination: A History of Phytochrome Research*. Academic Press, San Diego.

Sakamoto, M., Yazaki, N., Katsushima, N., Mizuta, K., Suzuki, H. & Numazaki, Y. (1995) Longitudinal investigation of epidemiologic feature of adenovirus infections in acute respiratory illnesses among children in Yamagata, Japan (1986–1991). *Tohoku J Exp Med*, **175**, 185–93.

Sankila, R., Joensuu, H., Pukkala, E. & Toikkanen, S. (1993) Does the month of diagnosis affect survival of cancer patients? *Br J Cancer*, **67**, 838–41.

Sapolsky, R. M. (2003) Taming stress. *Sci Am*, **289**(3), 86–95.

Sapolsky, R. M. (2004) *Why Zebras Don't Get Ulcers*. Owl Books, New York.

Sauman, I., Briscoe, A. D., Zhu, H., Shi, D., Froy, O., Stalleicken, J., Yuan, Q., Casselman, A. & Reppert, S. M. (2005) Connecting the navigational clock to sun compass input in monarch butterfly brain. *Neuron*, **46**, 457–67.

Saunders, D. S. (2005) Erwin Bünning and Tony Lees, two giants of chronobiology, and the problem of time measurement in insect photoperiodism. *J Insect Physiol*, **51**, 599–608.

Saunders, D. S. & Cymborowski, B. (1996) Removal of optic lobes of adult blow flies (*Calliphora vicina*) leaves photoperiodic induction of larval diapause intact. *J Insect Physiol*, **42**, 807–11.

Saunders, D. S., Vafopoulou, X., Lewis, R. D. & Steel, C. G. (2002) *Insect Clocks*. Elsevier, Amsterdam.

Schäfer, E. A. (1907) On the incidence of daylight as a determining factor in bird-migration. *Nature*, **77**, 159–63.

Schneps, M. H. & Sadler, P. M. (1987) *A Private Universe*. Annenberg/CPB, Harvard-Smithsonian Center for Astrophysics, Science Education Department, Science Media Group, Washington DC.

Searle, I. & Coupland, G. (2004) Induction of flowering by seasonal changes in photoperiod. *EMBO J*, **23**, 1217–22.

Sher, L. (2004) Alcoholism and seasonal affective disorder. *Compr Psychiatry*, **45**, 51–56.

Shiga, S. & Numata, H. (2007) Neuroanatomical approaches to the study of insect photoperiodism. *Photochem Photobiol*, **83**, 76–86.

Shimizu, I., Yamakawa, Y., Shimazaki, Y. & Iwasa, T. (2001) Molecular cloning of *Bombyx* cerebral opsin (Boceropsin) and cellular localization of its expression in the silkworm brain. *Biochem Biophys Res Commun*, **287**, 27–34.

Sinclair, B. J., Worland, M. R. & Wharton, D. A. (1999) Ice nucleation and freezing tolerance in New Zealand alpine and lowland weta, *Hemideina* spp. (Orthoptera; Stenopelmatidae). *Physiol Entomol*, **24**, 56–63.

Skala, J. A. & Freedland, K. E. (2004) Death takes a raincheck. *Psychosom Med*, **66**, 382–86.

Slater, P., Rosenblatt, J., Snowdon, C., Roper, T. & Naguib, M., eds. (2003) *Advances in the Study of Behaviour*. Academic Press, London.

Slingo, J. M., Challinor, A. J., Hoskins, B. J. & Wheeler, T. R. (2005) Introduction: food crops in a changing climate. *Phil Trans R Soc B*, **360**, 1983–89.

Smith, J. (2005) Starry starry nights. *Ecologist*, **35**, 56–62.

Smith, J. M. & Springett, V. H. (1979) Atopic disease and month of birth. *Clin Allergy*, **9**, 153–57.

Smits, L. J., Van Poppel, F. W., Verduin, J. A., Jongbloet, P. H., Straatman, H. & Zielhuis, G. A. (1997) Is fecundability associated with month of birth? An analysis of 19th and early 20th century family reconstitution data from The Netherlands. *Hum Reprod*, **12**, 2572–78.

Sobel, E., Zhang, Z. X., Alter, M., Lai, S. M., Davanipour, Z., Friday, G., McCoy, R., Isack, T. & Levitt, L. (1987) Stroke in the Lehigh Valley: seasonal variation in incidence rates. *Stroke*, **18**, 38–42.

Spears, T. (2004) Science ties illnesses to your month of birth: March is the cruellest month, with a long list of diseases. In *The Ottawa Citizen*, Ottawa, 4 February 2004.

Stefansson, H., Helgason, A., Thorleifsson, G., Steinthorsdottir, V., Masson, G., Barnard, J., Baker, A., Jonasdottir, A., Ingason, A., Gudnadottir, V. G. *et al.* (2005) A common inversion under selection in Europeans. *Nat Genet*, **37**, 129–37.

Stefansson, V. & McCaskil, E. I. (1938) The three voyages of Martin Frobisher in search of a passage to Cathay and India by the north-west, A.D. 1576–8. Argonaut, London.

Strode, P. K. (2003) Implications of climate change for North American woodwarblers (*Parulidae*). *Global Change Biol*, **9**, 1137–44r.

Subak, S. (1999) Seasonal pattern of human mortality. In Cannell, M. G. R., Palutikof, J. P. & Sparks, T. H. (eds) *Indicators of Climate Change in the UK*. Centre for Ecology and Hydrology, Huntingdon, 40–41.

Sung, S. & Amasino, R. M. (2004) Vernalization and epigenetics: how plants remember winter. *Curr Opin Plant Biol*, **7**, 4–10.

Swan, N. (2005) Prevalence of schizophrenia worldwide. In *The Health Report* (Australian Broadcasting Company). <http://www.abc.net.au/rn/talks/8.30/helthrpt/stories/s1358445.htm>

Tacitus (1971) *Annals of Imperial Rome* (translated by Michael Grant). Penguin Books, London.

Tang, D., Santella, R. M., Blackwood, A. M., Young, T. L., Mayer, J., Jaretzki, A., Grantham, S., Tsai, W. Y. & Perera, F. P. (1995) A molecular epidemiological case-control study of lung cancer. *Cancer Epidemiol Biomarkers Prev*, **4**, 341–46.

Tattersall, I. (2004) *Becoming Human: Evolution and Human Uniqueness*. Oxford University Press, Oxford.

Tauber, E. & Kyriacou, B. P. (2001) Insect photoperiodism and circadian clocks: models and mechanisms. *J Biol Rhythms*, **16**, 381–90.

Tauber, M., Tauber, C. & Masaki, S. (1986) *Seasonal Adaptations in Insects*. Oxford University Press, Oxford.

Taylor, R., Lewis, M. & Powles, J. (1998) The Australian mortality decline: all-cause mortality 1788–1990. *Aust N Z J Public Health*, **22**, 27–36.

Thomas, D. W., Blondel, J., Perret, P., Lambrechts, M. M. & Speakman, J. R. (2001) Energetic and fitness costs of mismatching resource supply and demand in seasonally breeding birds. *Science*, **291**, 2598–2600.

Thomashow, M. F. (1999) Plant cold acclimation: freezing tolerance genes and regulatory mechanisms. *Annu Rev Plant Physiol Plant Mol Biol*, **50**, 571–99.

Thoreau, H. D. (1906) In Torrey, B. (ed.) *The Writings of Henry David Thoreau* vol. V, Journals (March 5, 1853 to November 10, 1853) Entry for 23 August 1853. Riverside, Cambridge, Massachusetts.

Torr, S. J. & Hargrove, J. W. (1999) Behaviour of tsetse (*Diptera: Glossinidae*) during the hot season in Zimbabwe: the interaction of micro-climate and reproductive status. *Bull Entomol Res*, **89**, 365–79.

Torrey, E. F., Miller, J., Rawlings, R. & Yolken, R. H. (1997) Seasonality of births in schizophrenia and bipolar disorder: a review of the literature. *Schizophr Res*, **28**, 1–38.

Torrey, E. F., Miller, J., Rawlings, R. & Yolken, R. H. (2000) Seasonal birth patterns of neurological disorders. *Neuroepidemiology*, **19**, 177–85.

Tournois, J. (1912) Influence de la lumière sur la florasion du houblon japonais et du chauvre. *C R Acad Sci Paris*, **155**, 297–300.

Ueda, H. R. (2006) Systems biology flowering in the plant clock field. *Mol Syst Biol*, **2**, 60.

Vahidi, A., Soleimani, S. M. K. M., Arjmand, M. H. A., Aflatoonian, A., Karimzadeh, M. A. & Kermaninejhad, A. (2004) The relationship between seasonal variability and pregnancy rates in women undergoing assisted reproductive technique. *Iranian J Reprod Med*, **2**, 82–86.

Valverde, F., Mouradov, A., Soppe, W., Ravenscroft, D., Samach, A. & Coupland, G. (2004) Photoreceptor regulation of CONSTANS protein in photoperiodic flowering. *Science*, **303**, 1003–06.

VanderLeest, H. T., Houben, T., Michel, S., Deboer, T., Albus, H., Vansteensel, M. J., Block, G. D. & Meijer, J. H. (2007) Seasonal encoding by the circadian pacemaker of the SCN. *Curr Biol*, **17**, 468–73.

Vignaud, P., Duringer, P., Mackaye, H. T., Likius, A., Blondel, C., Boisserie, J. R., De Bonis, L., Eisenmann, V., Etienne, M. E., Geraads, D. *et al.* (2002) Geology and palaeontology of the Upper Miocene Toros-Menalla hominid locality, Chad. *Nature*, **418**, 152–55.

Visser, M. E. & Both, C. (2005) Shifts in phenology due to global climate change: the need for a yardstick. *Proc R Soc B*, **272**, 2561–69.

Visser, M. E. & Holleman, L. J. (2001) Warmer springs disrupt the synchrony of oak and winter moth phenology. *Proc R Soc B*, **268**, 289–94.

Visser, M. E., van Noordwijk, A. J., Tinbergen, J. M. & Lessells, C. M. (1998) Warmer springs lead to mistimed reproduction in great tits (*Parus major*). *Proc R Soc B*, **265**, 1867–70.

Vondrasová, D., Hájek, I. & Illnerová, H. (1997) Exposure to long summer days affects the human melatonin and cortisol rhythms. *Brain Res*, **759**, 166–70.

Wang, H., Sekine, M., Chen, X. & Kagamimori, S. (2002) A study of weekly and seasonal variation of stroke onset. *Int J Biometeorol*, **47**, 13–20.

Ward, R. (1971) *The Living Clocks*. Knopf, New York.

Wayne, N. L., Malpaux, B. & Karsch, F. J. (1988) How does melatonin code for day length in the ewe: duration of nocturnal melatonin release or coincidence of melatonin with a light-entrained sensitive period? *Biol Reprod*, **39**, 66–75.

Weale, R. (1993) Is the season of birth a risk factor in glaucoma? *Br J Ophthalmol*, **77**, 214–17.

Wehr, T. A. (1989) *Seasonal Affective Disorder: A Historical Overview*. Guilford Press, New York.

Wehr, T. A. (2001) Photoperiodism in humans and other primates. *J Biol Rhythms*, **16**, 348–64.

Wikelski, M., Martin, L. B., Scheuerlein, A., Robinson, M. T., Robinson, N. D., Helm, B., Hau, M. & Gwinner, E. (2008) Avian circannual clocks: adaptive significance and possible involvement of energy turnover in their proximate control. *Phil Trans R Soc B*, **363**, 411–23.

Wilkinson, P., Pattenden, S., Armstrong, B., Fletcher, A., Kovats, R. S., Mangtani, P. & McMichael, A. J. (2004) Vulnerability to winter mortality in elderly people in Britain: population based study. *BMJ*, **329**, 647.

Willer, C. J., Dyment, D. A., Risch, N. J., Sadovnick, A. D. & Ebers, G. C. (2003) Twin concordance and sibling recurrence rates in multiple sclerosis. *Proc Natl Acad Sci USA*, **100**, 12877–82.

Willer, C. J., Dyment, D. A., Sadovnick, A. D., Rothwell, P. M., Murray, T. J. & Ebers, G. C. (2005) Timing of birth and risk of multiple sclerosis: population based study. *BMJ*, **330**, 120.

Willmer, E. N. & Brunet, P. C. J. (1985) John Randal Baker. *Biogr Mems Fell R Soc*, 31–63.

Wiltschko, R. & Wiltschko, W. (1981) The development of sun compass orientation in young homing pigeons. *Behav Ecol Sociobiol*, **9**, 135–41.

Wirz-Justice, A. (1998) Beginning to see the light. *Arch Gen Psychiatry*, **55**, 861–62.

Wirz-Justice, A. (2005) Chronobiological strategies for unmet needs in the treatment of depression. *Medicographia*, **27**, 223–27.

Wright, K. (2002) Times of our lives. *Sci Am*, September.

Wuethrich, B. (2000) Ecology. How climate change alters rhythms of the wild. *Science*, **287**, 793, 795.

Yanovsky, M. J. & Kay, S. A. (2003) Living by the calendar: how plants know when to flower. *Nat Rev Mol Cell Biol*, **4**, 265–75.

Yasuo, S., Watanabe, M., Okabayashi, N., Ebihara, S. & Yoshimura, T. (2003) Circadian clock genes and photoperiodism: Comprehensive analysis of clock gene expression in the mediobasal hypothalamus, the suprachiasmatic nucleus, and the pineal gland of Japanese Quail under various light schedules. *Endocrinology*, **144**, 3742–48.

Yodzis, P. (2000) Diffuse effects in food webs. *Ecology*, **81**, 261–66.

Yoshimura, T., Yasuo, S., Watanabe, M., Iigo, M., Yamamura, T., Hirunagi, K. & Ebihara, S. (2003) Light-induced hormone conversion of $T_4$ to $T_3$ regulates photoperiodic response of gonads in birds. *Nature*, **426**, 178–81.

Yuen, J., Ekbom, A., Trichopoulos, D., Hsieh, C. C. & Adami, H. O. (1994) Season of birth and breast cancer risk in Sweden. *Br J Cancer*, **70**, 564–68.

Zhang, J., Schurr, U. & Davies, W. J. (1987) Control of stomatal behaviour by abscisic acid which apparently originates in the roots. *Journal of Experimental Botany*, **38**, 1174–81.

Zimmermann, T. (2003) Tempus Fugit: Prehistoric and Early Historic Devices for Telling Time. *Newsletter, Department of Archaeology and History of Art, Bilkent University*, **2**, 16.

Zucs, P., Buchholz, U., Haas, W. & Uphoff, H. (2005) Influenza associated excess mortality in Germany, 1985–2001. *Emerg Themes Epidemiol*, **2**, 6.

# INDEX